The Wired Nation

the text of this book is printed
on 100% recycled paper

The Wired Nation

Cable TV: The Electronic Communications Highway

RALPH LEE SMITH

HARPER COLOPHON BOOKS

Harper & Row, Publishers

New York, Evanston, San Francisco, London

THE WIRED NATION originally appeared in a shorter form as a special issue of *The Nation,* May 18, 1970.

Printed in the United States of America. For information address Harper & Row, Publishers, Inc., 10 East 53rd Street, New York, N.Y. 10022. Published simultaneously in Canada by Fitzhenry & Whiteside Limited, Toronto.

FIRST EDITION: HARPER COLOPHON BOOKS, 1972

LIBRARY OF CONGRESS CATALOG CARD NUMBER: 74–188032

STANDARD BOOK NUMBER: 06–090243–4

Designed by Yvette A. Vogel

to Mary Louise

CONTENTS

PREFATORY NOTE

The Wired Nation originally appeared in shorter form as a special issue of *The Nation,* May 18, 1970. I am grateful to *The Nation* for their permission to republish in this book much of the material that appeared in that issue.

So many people have helped me to understand cable TV and broadband communications that I cannot really make a proper list of my obligations. I will simply list three of special importance. Mr. Sidney Dean, Jr., chairman of the communications committee of Americans for Democratic Action and chairman of the cable committee of the American Civil Liberties Union, freely shared his extensive knowledge and his remarkable files with me when I was writing the original article for *The Nation.* Mr. Carey McWilliams and Mr. Robert Hatch, respectively editor and executive editor of *The Nation,* were immensely helpful in structuring and shaping the material. And the Fund for Investigative Journalism made everything possible by providing the grant on which the original version of *The Wired Nation* was written.

R. L. S.

The Wired Nation

1 *What Is Cable TV?*

I recently asked a high-ranking aide to one of the seven Federal Communications Commissioners: "What do you think the Commission should be doing about cable TV?" Despite the fact that the Commission has been issuing rules on cable for a number of years, and was in the midst of a big new rulemaking proceeding on cable when I interviewed him, he sighed and waved his hand. "I don't know what we should be doing. If you get any ideas, let me know!" I soon learned that, on many or most of the basic issues posed by cable, the Commissioner is in the same quandary.

Cable television was originally called community antenna television and is still widely known as CATV. It began as a minor adjunct to the present system of over-the-air commercial broadcasting. Now often called broadband communications, it is on the verge of becoming a major medium in its own right. As cable systems are installed in major cities and metropolitan areas, the stage is being set for a communications revolution—a revolution that some experts call "The Wired Nation."

In addition to the telephone and to the radio and television programs now available, cable can bring into homes and into business places audio, video and facsimile transmissions that can provide newspapers, mail service, banking and shopping facilities, data from libraries and other storage centers, school curricula, complete course offerings leading to college degrees, cultural programs, community-originated TV fare, and other forms of information too numerous to specify. In short, every home and office can contain a communications center of a breadth and flexibility to influence every aspect of private and community life.

But despite the importance and imminence of this new force in society, its possibilities and problems are still little known to the public. Few local, state or national legislators are better off. Even members of the Federal Communications Commission, of Congressional committees concerned with communications, and of state and municipal regulatory bodies are not sure what they should do. The problem is made more difficult by the fact that much of the information on cable television that government bodies and the public do receive, comes from sources that have their own interests to serve. It seems, therefore, that a basic primer on cable TV would be of value to both legislators and the public, and it is hoped that this book will serve that purpose.

Cable television, or CATV, is a system whereby the set owner receives his signals through a coaxial cable instead of over the air. Persons who wish to enjoy the service pay an installation charge to have their set hooked to the cable, and a monthly service charge.

If it is true, as it seems to be, that cable TV is about to effect a revolution in communications, it must be said that a revolution has rarely been created by persons of less revolutionary intent. The original purpose, and still the principal function, of CATV is to provide the viewer with better reception and a larger selection of existing TV stations than he can get from the air.

Cable TV was invented and developed by TV servicemen in small towns and rural areas, where reception is poor or nonexistent, and where in any case very few channels are on the air. The Federal Communications Commission has never required broadcasters to set up relay systems to provide adequate TV service to the substantial portions of the nation that it does not suit the commercial interests of the industry to serve.

To remedy this situation, Robert J. Tarlton of Lansford, Pa., twenty years ago set up the world's first commercial cable TV system. Born in Lansford in 1914, Tarlton graduated from high school in 1933 and opened a radio sales and service shop. During World War II he served as an enlisted communications technician in Italy, then returned to his shop.

When TV sets became commercially available in the late 1940s, Tarlton had trouble selling them because reception in Lansford was abominable. The nearest stations were in Philadelphia, sixty-five miles away, the signals reaching Lansford were very weak, and further blocked by a mountain that overshadows the town. Tarlton experimented in 1949 with installing individual antennas for set owners on the mountain. That worked fairly well, and Tarlton quickly got a better idea. It was, however, beyond his ability to finance, so he and several friends pooled their resources and set up a firm called Panther Valley Television Company.

Panther Valley built a tall master antenna atop the mountain, to spear the faint Philadelphia signals. These were fed into an amplifier to bring them back to full strength, and then into a coaxial cable that was strung on poles down the mountainside and into town. The company offered to hook customers up to the cable for an installation charge of $125 and a monthly service charge of $3. Television-hungry residents of Lansford immediately began buying sets from Tarlton's shop and "going on the cable." They received three Philadelphia channels with greater fidelity and clarity than did a lot of people living within ten miles of Philadelphia—or even in the city itself.

Today, with its system modernized and rebuilt, Panther Valley Television provides twelve channels to 2,900 residents of the Pennsylvania hill towns of Lansford, Coaldale, Havto and Lake Havto, who would otherwise have little or no TV. Tarlton remains president of the company, and also of Titusville Cable TV in Titusville, Pa. He has an assured niche in history, in the special catalogue of persons who, without intending to, Really Started Something.

A modern cable system is simply a more sophisticated version of Tarlton's original rig. A tower is constructed on a hill or other spot selected for good reception. The antenna system on the tower is carefully engineered, usually with a separate antenna for each channel that is to be received. In some instances, distant signals come to the tower by a relay system of one or more microwave transmitters.

At the foot of the tower is a small control station, called the "headend," where the signals are brought up to maximum strength and clarity. Here, also, some of the signals may be rechanneled. It is customary, for example, for cable systems to put UHF stations on empty VHF channels. Amplifiers placed at distances of 1,500 to 2,000 feet along the trunk line into town to keep the signals strong. From the main cable, feeder lines, "tap-offs," and "housedrops" carry the signals to individual streets and the subscribers' homes. These days, the installation fee usually runs from $10 to $20, and the monthly service charge is about $5.

Cable TV has proved to be immensely popular. By June 1971, over 2,750 systems were operating in the United States, and there were about 2,400 additional communities where CATV permits had been issued but construction had not yet started. In recent years the number of subscribers has grown at a rate of 20 to 25 percent annually, and the systems now serve about nineteen million viewers—about 9 percent of the U.S. TV audience. Investment in plants exceeds $500 million and is soaring; estimated annual revenues of the industry are about $350 million and are also rising fast. Industry leaders estimate that by 1980 half the nation will be on the cable, and that the industry will have a net worth of $5 billion and annual revenues of over $2 billion.

In the first stage of the CATV boom, most of the activity has been in those cities, towns and communities that do not qualify as heavily concentrated markets, and which have therefore not been regarded by the FCC or the broadcasters as deserving full TV service. About 46 percent of the population lives in towns and cities ranging from 2,500 to 50,000 population, and few of these communities have their own TV stations. Today, many such communities get more and better TV than do residents of major market areas. Towns in northern Vermont receive New York City channels more clearly than many residents of the five boroughs, whose reception can be fogged and snowed by tall buildings. In Aspen, Colo., residents could receive only one channel before the cable arrived. Now they get all three major networks plus an independent station and a channel of educational TV. Fort Madison, Ia., gets seven channels plus weather information on an eighth and a view of an AP wire service ticker on a ninth. Bradford, Pa., which formerly considered itself lucky to receive one channel, now gets eleven.

Most cable systems are still small; the average setup has about 2,150 subscribers, with a few of the largest systems serving about 45,000, and a number of small ones serving fewer than 100. Restrictive policies of the Federal Communications Commission, developed in response to urgent pressures from the broadcasting industry, have retarded the growth of CATV systems in metropolitan areas. Cable systems are nevertheless starting to come to the nation's cities, and it is here that the big growth stage for CATV will occur in the 1970s, and where the great promise of the cable will be fulfilled—or missed.

Clear reception and a wide selection of stations are of course desirable, and the success of cable TV shows how badly they are wanted. But the excitement about the Wired Nation is not being generated by these present major uses alone. "Cable television," says a report issued in 1968 by Mayor Lindsay's Advisory Task Force on CATV and Telecommunications in New York City, "offers the most promising solution to a number of difficult communications problems facing our larger cities." Testifying before the Federal Communications Commission during cable rulemaking proceedings in 1971, John W. Macy, Jr., President of the Corporation for Public Broadcasting, was even more enthusiastic. "These hearings," he said, "are less about the future of a particular form of communications technology than they are about America's future." James Day, President of Educational Broadcasting Corporation, testifying at the same hearings, called cable "a potential public service of almost incalculable proportions." These and other advocates of cable are referring to *potential* uses to which the cable can be put. To understand them, one must look first at the star performer in this drama—the cable itself.

A coaxial cable consists of three elements. In the center, like the lead in a pencil, runs a copper wire. This is surrounded by an insulating layer of polyethylene foam, which accounts for the major portion of the diameter of the cable in cross section. The insulation, in turn, is surrounded by a tubular shield of braided copper wire or a seamless aluminum sheath.

When a current or signal is introduced into the cable, an electromagnetic interaction takes place between the center wire and the surrounding sheath. The interaction prevents currents from radiating off the cable. This is the secret of the cable's key characteristic—its immense capacity for carrying electronic signals, data and information. Comparing a coaxial cable to a telephone wire, says FCC Commissioner Nicholas Johnson, is like "comparing Niagara Falls to a garden hose." As we shall see, the phone company is very much aware of the comparison.

Television is a colossal hog of the electronic frequencies. The elbowroom required by each channel is what makes the over-the-air very high frequency (VHF) TV spectrum the scarcest of our natural resources. The Federal Communications Commission permits only twelve channels of TV to be broadcast over this choicest part of the spectrum, allowing each one a bandwidth of six megahertz (hertz means cycles; megahertz is usually abbreviated Mhz). By way of comparison, six megahertz of bandwidth can accommodate 640 AM radio signals or 1,200 voice telephone channels. When additional allowance is made to avoid overlapping and interference, and for the further restrictions imposed by the economics of a commercially based broadcasting system, it works out that nearly half of the U.S. TV audience cannot receive more than six channels; many receive less; and the picture quality is inferior on many of the channels that can be received.

One of cable TV's great potentials is its inherent ability to end this economy of scarcity, on which the power of the present TV broadcasting oligarchy is solidly based. With existing equipment and technology, a single coaxial cable can carry between 28 and 35 channels of television, plus the entire AM and FM radio bands and a quantity of other nonvisual electronic signals. If more channels are wanted, a second cable can be laid, and a third, and a fourth. And, as technology produces increasingly sophisticated input equipment and amplifiers for cable communications, each cable will be able to carry more signals. Nathaniel Feldman, a Rand Corporation engineer, predicts that in ten to twenty years a four-cable system will be able to carry 400 channels of television.

In addition to opening up as many channels for TV broadcasting as any community could possibly use, cable TV greatly facilitates more diversified use of television by cutting the cost of transmission. In 1969, President Nixon, under sharp prodding from Congressman Tobert H. Macdonald (D., Mass.), chairman of the House Commerce Committee's Subcommittee on Communications, finally released the text of a Johnson administration Task Force report on national communications policy. Among other things, the report concluded that money, even more than lack of space on the spectrum, was a major barrier to the expanded use of TV. The cost of building and running over-the-air transmitters, and programming costs, which are rising at a rate of about 8 percent annually, make any expansion of the present system almost prohibitively costly.

In a CATV system, it is possible to transmit directly over the cable without receiving any signal from the air. Some cable systems already do this on one or more empty channels; they call it "cablecasting" or "program origination." This eliminates the high cost of building and running an over-the-air transmitter. In addition, while a commercial over-the-air broadcaster derives his entire revenues from his programming, a cable system does not make its money on what it transmits. The cable runs on the profit from the subscribers' fees, whether or not the system is used for cablecasting.

Together, then, the elimination of channel scarcity and the sharp reduction of broadcasting cost, can break the hold on the nation's television fare now exercised by a small commercial oligarchy. Television can become far more flexible, far more democratic, far more diversified in content, and far more responsive to the full range of pressing needs in today's cities, neighborhoods, towns and communities.

A second area of great promise was made vivid by an exhibit at the 1968 Annual Convention of the National Cable Television Association. It displayed a home communications center, in which the user, through appropriate switching circuits, could enter into two-way exchanges with local stores, could "dial-a-play," and could have at his finger tips the full information contained in vast libraries. This is no dream. The cable could carry it all, and the technology is in existence or soon will be.

Looking a bit farther into the future, it may be that we are heading toward a single, unified system of electronic communications. "Before very long," says Brenda Maddox in a booklet entitled *Communications: The Next Revolution,* issued by the London *Economist,* "information theory will have been brought to its logical conclusion in public communications; there will be a single unified network for all kinds of messages . . . separate systems for telephones, telegraph, television and data transmission will disappear. Information will flow through the network as on-off digital signals and appear as pictures, sound or print, according to the choice of those sending and receiving it." The speed with which such a network could rattle off bundles of information is hard to appreciate. "Information," says AT&T's W. M. Ellinghaus, "can now be moved at the rate of 1.5 million words per minute. *Gone With the Wind* can be transmitted in about twenty seconds. The Bible would take a little longer—about half a minute."

2 *New Kinds of Television*

In the early years, cable TV operators did not originate any programming over their own wires. In time, however, to make the service more attractive, systems began filling one or more empty channels on their viewers' sets with such simple fare as weather reports, stock market quotations, and views of an AP or UPI news ticker. Such "programming" costs little, and is easily provided by even a very small system. Soon, a few systems went a step farther—they began to transmit live local material. Today, 5 to 10 percent of cable systems offer live programming of local origin, usually transmitted for a few hours a day, or irregularly when events of interest take place. These include newscasts, religious programs, school activities, county fairs, fund-raising drives, sports, cultural events, political debates, public hearings, school board meetings, children's programs, and daily variety shows featuring local persons and events.

In southwestern Massachusetts, High Fidelity Cable Television of Great Barrington serves some 2,800 subscribers in Great Barrington, Stockbridge, Lee and Lenox. Without the cable, many viewers in this area can get only one faint signal, from a VHF station in Schenectady. The cable brings in eleven VHF, UHF and educational stations from Hartford, New Britain, Springfield, Albany, Schenectady, New Haven and New York City.

None of this, however, provides the subscribers in the four towns with coverage of their own communities. The cable system has therefore hired a full-time news director, who broadcasts two local news roundups, complete with video-tape clips, at 6:15 and 11:15 each evening. Special events such as political campaigns and school activities are covered as they occur. Community interest is high; in Great Barrington, 95 percent of homes within reach of the cable subscribe to it. "We are an entirely new news medium in the southern Berkshires," says John Mooney, the system's general manager.

For a more elaborate program schedule, a system will often hire a full-time program director, and will hire local persons on a part-time basis to do everything else. Often, the part-time personnel are local youths and high school students, who receive training in nearly every aspect of TV program production while contributing a high degree of interest and enthusiasm.

Such an operation is that in Henderson, North Carolina. Henderson is located in the center of a triangle of larger cities —Spartanburg and Greenville, South Carolina and Asheville, North Carolina. The area as a whole is ranked 44th on the Federal Communications Commission's list of the nation's top one hundred markets. But Henderson is isolated from the three outlying cities by mountains, and is thereby cut out of both TV reception and TV coverage. "No commercial station could afford to locate here—the economy is too limited," says Robert W. Davidson, general manager of Cablevision of Henderson, the town's cable system, "so Cablevision is the closest thing to a hometown TV station for our 7,000 residents."

Davidson hired a program director and then took on a high

school student, a senior with an interest in radio, as a control room trainee. "He learned everything in three days!" Davidson says. Two more students were hired as cameramen, and three more to work the control board; they learned with equal speed.

Cablevision's programming day begins with a morning exercise program hosted by the system's secretary and Girl Friday, with Davidson and his systems engineer operating the cameras. The students come in at 4 PM and the place begins to jump. At 6 PM the system goes back on the air with a full half-hour of local news followed by an "Early Tonight Show," featuring area guests and local topics. Special events covered by the system have included local elections and videotaped cablecasts of the Henderson High School basketball team in the State championships. Davidson also arranged to procure and show the films of local high school games made by team coaches for use by the coaches and their teams. He pays a student to attend each game and take notes; then, as the films are shown on the cable, the student and one or more local coaches provide running commentary.

Soon the students were providing their own ideas and making their own programs. One was a series of ten thirty-minute shows called "Sights and Sounds of the 60s" put together by Davidson and the students from photographs, borrowed news footage, film clips, and over six hundred slides and several hundred feet of film shot by one of the students who is a photographer.

Local merchants have knocked on Davidson's door to buy moderately-priced advertising spots on the new medium, which enables them to reach local residents via TV for the first time. Already, advertising revenue from local sources often runs greater than income from either installation or subscription fees.

Another kind of communications void is that suffered by

suburban communities clustering around large cities. Such a community is Wayne, New Jersey, a suburb of New York City and Newark, with a population of 50,000, which has never had a TV station of its own. Neither do any of the six smaller towns surrounding Wayne—Ringwood, Wanaque, Bloomingdale, Butler, Pompton Lakes, and Oakland. In addition, none of these towns has a daily newspaper. From the standpoint of communication they depend largely on the TV stations of New York and Newark, which rarely give them any coverage.

Recently a cable television system was installed, covering Wayne and the six other towns. In addition to retransmitting a selection of over-the-air stations to its subscribers, the cable system built a small studio, and early this spring began transmitting thirty hours a week of programming over an empty channel on subscribers' sets. Local news in all seven towns is covered by a full-time newsman who roves the communities with a half-inch portable tape recorder. Parades, festivals, sporting events, meetings of civic and community groups, and other local activities are covered live, on tape, or by the newsman in his daily news broadcast.

In addition, local issues are aired in discussion programs, with one or more local guests in the studio and with an open phone line for two-way dialogue with listeners. Mayors of the towns report that understanding of, and interest and participation in, local problems has increased. At election time all candidates receive air time over the cable channel, and on election night coverage of the local races is thorough.

William Paterson College is located in the cable system's area, and this college, like many institutions of higher education, has TV studio facilities of its own. The college studio has been hooked up to the cable system, and student-created programs are broadcast to all subscribers. Arrangements are now being completed to broaden the college's entire relationship to the surrounding communities by broadcasting educational programs and courses from the college studio over the cable, at times when adults as well as students can view them, for their intrinsic interest and/or as part of decentralized and off-campus programs leading to degrees.

The cable system is now working on another project—continuing education programs for physicians only, transmitted to them on a "scrambled" cable channel which only they can unscramble. Doctors usually find it difficult to leave their practices to attend refresher and updating programs at universities or medical centers. The cable system therefore plans to bring the material to the doctor. When the project has been launched, the cable system will offer the programming to other cable systems in New York and New Jersey by microwave link, creating a miniature network of programming for the professional use of doctors only.

These local initiatives emphasize a second basic shortcoming in the present system of over-the-air TV broadcasting. In addition to discriminating against the large percentage of the population beyond the clear signal range of a few major metropolitan centers, it has not been able to offer real community service or community expression. When over-the-air broadcasters say that they provide community or local signal, they mean that they send out a highly generalized service to a large, poorly defined audience.

The precise coverage of a cable TV system makes it the logical medium for solving the problem of local television service. "This potential," said former FCC Commissioner Kenneth A. Cox in a speech to the National Association of Broadcasters, "excites a good many social scientists, who envisage every little town having its own outlet for local self-expression, even though it is far too small to support a regular broadcast station in the foreseeable future. Indeed, these people believe that a cable system in metropolitan Washington or in New York City would be able to provide pinpointed service to subcommunities within the larger complex, either by franchising several operations along political or ethnic or by other relevant lines, or by equipping an overall system with switching or filtering devices to enable the operator to deliver certain programming only to those areas likely to be interested in it."

Most CATV systems provide free or at-cost service to schools and municipal offices. Trenton, N.J., made this a requirement for the franchise. In Caspar, Wyo., the system shares its facilities and studio with the school district. Scheduled course material and films are transmitted directly to TV sets in the classrooms. In summer, review courses can be seen at home by students needing remedial instruction. The school system plans to add a high school news program, a teen discussion forum, a disc jockey show, and a job and career information program for young people.

Similar arrangements have been worked out between the cable system and the school system in Grand Junction, Colorado. The schools there and the cable operator are also investigating the use of a sub-channel on the cable for selective communication among schools, teachers and students, by means of a two-way switching system. In Overland Park, Kansas, TeleCable Corporation has begun to test a two-way cable operation which will enable homebound students to participate in public school classroom activities. The system requires a normal TV receiver tied into the cable system and a small television camera in the home, and a special terminal in both the child's home and the classroom. A disabled or other homebound child can watch and listen to the class in progress, and can "raise his hand" by pressing a button on his terminal, which registers on the terminal in the classroom. The teacher answers by pressing a button. The two can then converse, and the teacher can see the child on the TV screen in the classroom. The system is designed so that one teacher can handle many children. Overland Park has more than two hundred children who are unable to attend school regularly because of some disability, and there are two million such children nationally.

Such applications are only a bare beginning of the cable's potential. For example, six TV channels can be carved out of the electromagnetic spectrum at wave lengths below channel 2. These channels are not used for over-the-air broadcasting because they have inferior carrying qualities, but they can be carried on the cable with no significant loss of signal strength or clarity.

Similarly, there is a gap between channels 6 and 7 on standard receivers, big enough to accommodate nine TV channels.* Some or all of these channels could be devoted to municipal use. One or more of them could be programmed by the city's health and social service departments, and a special channel could perhaps be used by these departments to serve low-income areas. Community groups, and local talent and culture, could also find expression over the cable's facilities. There is enough channel capacity available in the cable to provide ample room for all the fare that they could possibly wish to present.

In depressed areas and ghetto neighborhoods, cable TV could perform functions of special importance. For example, in a report on the Watts riots, the RAND Corporation recommended that $750,000 be invested in a local, over-the-air commercial UHF facility for the community. But a cable system could provide Watts with far more precise coverage at lower

* The gap runs from 88 MHz to 174 MHz. The frequencies between 88 MHz and 108 MHz are used for the FM broadcast band. The 108–120 MHz band is for aircraft navigation, and must be kept clear of all radiation danger, even from the almost completely radiation-free cable. But this leaves the remainder of the "gap," from 120 to 174 MHz, which, for technical reasons, are avoided by over-the-air broadcasting, but can be used without difficulty by the cable.

cost. The same is true of other city neighborhoods that now live in anonymous invisibility until they are torn by riot.

The cable could affect the cultural content of TV in many ways. Large budgets for national production of over-the-air public television, while a valuable and important expansion of the uses of TV, may not be the real key to opening up the medium's full artistic and cultural potential. The advocates of public television, says Leonard Chazen, an assistant professor of law at Rutgers and a consultant to President Johnson's Task Force on Communications Policy, "believe in an identifiable commodity known as 'good programming' that can safely be entrusted to a single well-motivated organization. But with all we have learned about the elusiveness of artistic standards, can we really be convinced that centralized decision making, as represented by a public network, is the way to make television come alive? Certainly this is not the answer suggested by experience in other communications media, and in particular, the movies." University-based audiences, low-budget filming methods, and the proliferation of independent producers, distributors and backers, all gave the creative script writer and film maker somewhere to turn besides the few Hollywood giants. The proliferation of outlets and reduction of costs inherent in cable TV, says Chazen, could have a similarly creative effect.

In some areas local programming is already producing material of artistic merit that would otherwise never have found its way onto TV. In Manhattan, Robert Bleyer, program manager of TelePrompTer Manhattan Cable TV, a system franchised in Manhattan, received a visit from Al Fann, the founder and director of an amateur street theater company in East Harlem. Fann had written a play called "King Heroin," which his group had performed in community centers, schools, and churches in Harlem, and which it would soon perform at Marymount College in the Northern suburbs of New York.

Would TelePrompTer be interested, Fann asked, in making a videotape of the play for possible showing on the cable?

Bleyer sent an assistant to Marymount College on the afternoon of the scheduled evening performance there, and the assistant videotaped two-thirds of the play on the college's bare stage during final rehearsal. He then had to leave in order that the props and stage could be set up, so a few days later the rest of the play was acted and taped in TelePrompTer's studio, which occupies the former laundryroom of a large apartment building on Manhattan's 178th Street. The two sections were spliced together, minor editing was done, and the play was shown on TelePompTer's origination channel in the Fall of 1970. Most reviewers who saw it were overwhelmed. "This week," Jack Gould, television critic for the *New York Times,* wrote in the *Times* after seeing the tape, "marks a chapter in the evolution of communications. Electronics can and will expand the audience for truly community theater, in this instance a largely black company crying out against genocide of their children. [It] demonstrates anew that slick settings and fancy camera work are no substitute for authority, integrity, and a willingness to experiment. . . . 'King Heroin' stands as one form of tomorrow's TV: How much can be accomplished on a modest scale given the will and conviction."

Two cable systems have been franchised in Manhattan, with TelePrompTer covering the northern half of the island and Sterling Manhattan Cable TV covering the southern half. Both systems operate under franchises that require them to provide open channels for use by the general public. Any individual or group can have access to these channels without charge simply by applying to the company, an arrangement that dramatically signals the approaching transformation of TV from a medium based on channel scarcity to a medium with enough channels to meet all demands. The first public access channels were opened by the two systems on July 1,

1971, and at the same time a civic group called Open Channel was created to spread public knowledge of the facility and to promote its use. Things began slowly, but requests for time on the channels are now increasing. Users include civic, educational, and cultural groups, and groups interested in the emerging half-inch videotape and videocassette technology. Groups of the latter type include the Alternate Media Center of New York University, which has been working with Sterling to provide full coverage of such events as recent hearings on mental and public health problems in New York City prisons.

Another local and community service potential of cable television is so important that it must be separately discussed. CATV could arrest and reverse some ominous developments in American electoral politics.

Air time, and most especially TV time, has become an indispensable part of the electoral process. In June 1971 the Federal Communications Commission reported that in 1970, a nonpresidential election year, candidates for Congressional, state and local office spent $50.3 million on air time, an increase of 57 percent over the last comparable off-year figure of $32 million in 1966. Even for this amount most of the candidates received relatively little coverage. Full coverage would have cost them far more than their campaign chests could afford.

The indispensability of TV exposure, combined with its high cost, is probably the most important single factor in the rapidly increasing trend, for politics, at all levels, to become a game of the privileged, wealthy few. The cost is made even higher by the inexactness of over-the-air coverage. People living in the city see and hear radio and TV spots purchased by candidates in outlying suburban districts and communities. Only a fraction of those who will actually vote in the candidate's district are included in the metropolitan coverage, but he must pay for the station's entire audience.

Once the candidate is elected, another dimension of the problem becomes apparent. "It forces one to question whether government can ever realistically be expected to sustain a vigilant posture over an industry which controls the very access of government officials to the electorate," says FCC Commissioner Johnson. "The fact that roughly 60 percent of all campaign expenses go to radio and television time gives but a glimmer of the power of broadcasting in the lives of Senators and Congressmen."

The power of the broadcast lobby probably explains the long-standing refusal of Congress even to look at the problem. But in September 1969 a bipartisan group of Congressmen and Senators, with the backing of the National Committee for an Effective Congress, proposed legislation that would require broadcast stations to provide a certain amount of time at reduced cost to political candidates. The bill was ludicrously timid: it applied only to U.S. Senators and Representatives, and mandated no free air time.

In October 1969 the Senate Commerce Subcommittee on Communications held hearings on the bill. All the witnesses favored it except the broadcasters, who were unanimously opposed. Subsequently, several of them voluntarily offered reduced rates to candidates. Having profited for years from the democratic process, the broadcasters hope that this strategic concession will forestall Congressional action which, even in its proposed mild form, would establish federal authority and could lead to more comprehensive legislation.

Other opponents of the bill were incumbent Congressmen and Senators with substantial independent wealth or financial backing, who prefer to give to potential challengers as little air time as possible.

In 1971 President Nixon vetoed a bill that would place ceilings on political campaign expenditures. No matter what happens to these various proposals, enactment of legislation requiring adequate amounts of free time to all political candidates is obviously far in the future, if indeed it ever comes at all. And even such legislation would not alter two other problems of over-the-air broadcasting—that coverage bears no relationship to political geography, and that most U.S. communities do not even have an over-the-air TV broadcasting facility. All three problems, however, can be successfully dealt with through community-based cable systems.

At the national level, cable TV was first used extensively in 1968. Each major party prepared a video tape, giving a full presentation of its Presidential candidate and his views. These tapes were played on 350 cable systems, with a potential audience of 4.5 million persons, on time donated by the cable operators.

3 *The Coming of the Cable*

That, then, is the broad picture of the Wired Nation. What follows are battle reports of the forces and interests that are already besieging it.

At least two parties have been watching the unfolding drama of the Wired Nation with more than detached interest. One is the present commercial broadcasting industry, whose own interests—and survival—are directly threatened. The other is the telephone company, whose relationship to cable TV and the broadband communications revolution is problematic and complex. Both parties, however, decided that the way to be sure that *they* have a voice in the shape of things to come is to own the physical plant on which the revolution will be based. Moving swiftly and taking full advantage of the almost total lack of national policy on cable TV, they both began to build or acquire systems. Newspapers and other communications media also began to buy in, so that half the cable systems in the United States are now owned by other media and communications interests. If the trend is not stopped and reversed, the new technology will be largely the property of old interests that may find themselves with strong motives for abridging or not developing its full potential.

CATV is by its nature a monopoly and must secure franchises from local municipal governments. A scramble for franchises began as soon as the industry came into existence, and it continues to this day.

Five, ten, or more companies will often compete for the franchise in a community that promises a high percentage of potential cable subscribers, and big sums are being spent on lobbying and in legal battles. As CATV moves into increasingly large communities, says *TV Communications,* "franchise fights will become bloody and drawn out. . . . Cost of obtaining the CATV franchise for a major city may reach hundreds of thousands of dollars."

The companies are willing to make such expenditures because the prospects are, to put it mildly, intoxicating. In October 1968 the investment research firm of Drexel, Harriman, Ripley published a thirty-two-page "Industry Report on Community Antenna Television (CATV)." It presents the hypothetical case of a system constructed in a community of 30,000 to 40,000 people, of whom about 10,000 might subscribe. There are assumed to be 100 homes per square mile. Such a community would require 100 miles of cable, at an installed cost of $4,000 per mile, or a total of $400,000. A good antenna and headend installation would cost about $75,000, and legal fees, promotional expenses and other costs would come to about $85,000. The total investment in the rig would therefore be $560,000. In the example the system is to be depreciated over ten years, and the cable operator has borrowed $400,000 of the required capital at 7.5 percent interest.

When such a system has signed up 5,500 subscribers—55 percent of the potential total in the community, which many systems meet or exceed—it will be producing, after all costs have been met, depreciation is accounted for, and interest and taxes are paid, an annual profit of $167,000. At that rate it would more than repay its total cost in less than four years. "A system operator might reasonably expect to achieve a pretax margin of 30 to 35 percent," the study says coolly.

Still more money can be made by charging extra fees for extra services. "The addition of these services might conceivably mean a doubling or tripling of subscriber fees in time," says the report. Profits can also be increased by hiking the subscribers' basic monthly fee, and the report gives hints on how to get these raises past the local authorities.

If the operator ever got tired of this gold mine, he would have no trouble disposing of it. Such a system, says the report, often sells for $300 to $350 per subscriber, and the price can go as high as $400, $500 and even $700. If the hypothetical system were sold at $300 per subscriber, it would bring $1.6 million, almost three times its original cost. In practice, that often happens. The operator sets up a system, runs it for about five years while depreciating it rapidly, paying little or no taxes and pocketing hundreds of thousands of dollars. He sells it at an enormous profit to another person who is free to repeat the process.

The economic patterns of cable change as it moves into large cities and urban areas. It costs about $4,000 per mile to lay cable in smaller towns, but the cost can be more than $50,000 per mile in big cities. Investments in plant for urban systems are therefore high—an estimated $6 million to wire Columbus, Ohio, an estimated $20 million to wire St. Louis, and much more to bring the cable to New York, Chicago, Los Angeles. The number of people who will subscribe, the kinds of supplementary services that will be offered, and the charges that will be made for these services, are all uncertain. The payoffs on these large investments will therefore not come for seven or eight years after the investments have been made, and no one is really sure how much the payoffs will be. The fact remains, however, that when big cities announce their interest in granting cable franchise, there is no lack of bidders. At this writing, twenty parties are seeking franchises in Chicago. One of them, Pacific & Southern Broadcasting Company, headquartered in New York, said at the beginning of 1970 that it had already spent $209,157 on its Chicago franchise bid. Investors in the firm include more than thirty congressmen.

Obviously, the correct procedure for city fathers in this situation is to stand guard over the citizens' pocketbooks. Given the competition for franchises, city governments could insist on reasonable subscription fees, put a liberal but definite ceiling on profits and capital gains, and require a large array of services. Unfortunately, things have gone just the other way. The fox and the watchdog have compared notes and discovered that they have parallel interests in the chickens. Says a 1969 statement on cable TV issued by Americans for Democratic Action: "Public interest safeguards," including "reasonable pricing . . . are largely ignored." Many municipalities soak the cable operators, and then permit them to make unconscionable profits, in which the cost of the municipal shakedown can easily be absorbed.

In the early days, municipalities did not ask for a share of the take, but most of them now make it clear that anyone who hopes for a franchise must give the city its cut. Many agreements allow the city 5 percent of the gross, and some go higher. The franchise in Albuquerque, N.M., assigns the city 5 percent of the gross up to $250,000, 10 percent from $250,000 to $500,000, and so on up to 36 percent at $1.3 million. In Modesto, Calif., Cablecom General won the franchise by offering the city a $32,000 advance payment, and a sliding scale of the gross, running from 5 to 37 percent. Theta Cable is reported to have plunked down $845,000 for the privilege of wiring up the communities in the canyons surrounding Santa Monica.

It need hardly be said that many persons who got in on the ground floor of the business are now multimillionaires. In

1950 Milton Shapp started in cable TV with an investment of $500, and was soon involved in a number of cable enterprises. In 1958 he testified before the Senate Commerce Committee, and was questioned by Kenneth A. Cox, then the committee's counsel and subsequently an FCC Commissioner:

> MR. COX: Did you install the system in Clarksburg W. Va.?
> MR. SHAPP: Yes.
> MR. COX: What was the total cost of installation in that, approximately?
> MR. SHAPP: The cost of installation is different than the cost of invested capital. There is approximately $200,000 invested in Clarksburg, but the cost of the system is $350,000 to $400,000. They used much of the money collected from subscribers to expand the system.
> MR. COX: I have some place in my notes a figure of $975,000 in connection charges. Would that be correct?
> MR. SHAPP: It could be.
> MR. COX: This would be about three times the—
> MR. SHAPP: You have picked out one of the three most profitable community antenna systems in the United States.
> MR. COX: And the rates that you charge there are not, and never have been, regulated either by the FCC or by any agency of the state of West Virginia?
> MR. SHAPP: No.

It should be noted that the $975,000 was for installation charges alone, and did not include any of the monthly subscription fees. In 1966, when he ran for governor of Pennsylvania, Shapp sold his CATV interests for a reported $10 million.

Obviously, what is needed, and needed urgently, is a different and better kind of watchdog system.

In another area, that of programming, the display booths at the 1969 Convention of the National Cable Television Association were intriguing, but not invariably reassuring. At one of them, a pretty girl in the scantiest of bikinis sat on a high stool, wiggling her toes in the midst of a sea of suit-and-tie cablemen. She was a come-on for CATV Bingo. The promoter provided all the material and information for the use of an empty CATV channel to play bingo by cable. Local merchants distribute the cards to customers, who then rush home, glue themselves to the cable, and play solitaire bingo all day long, phoning the cable station when they win. "CATV Bingo is sweeping the country," the firm's literature said.

Less obviously stultifying is an ambitious cablecasting package developed by Gridtronics, Inc., a subsidiary of TeleVision Communications Corporation (TVC), run by Alfred Stern. Stern is one of the CATV industry's most eloquent spokesmen. "I submit that we must take the 'mass' out of the mass media when we talk about the future of television in America," he said in a speech to the 1968 NCTA Convention. "Cable communications is the only way available to the set owner in America to decide what he wants to see." Unfortunately, Stern's ambitious contribution to cablecasting seems designed to put "mass" into this new medium.

The Gridtronics cablecasting package consists of four channels of material that the cablecaster can put over his cable. They are described in the firm's literature:

> *Information Channel:* The dramatic presentation of history, literature, travel—to heighten an individual's perception of contemporary life.
>
> *Instructional Channel:* Films that teach at every level.
>
> *Startime Theatre Channel:* An uninterrupted entertainment series of blockbuster movie features.
>
> *Professional Channel:* A dynamic new medium providing vital information for the physician.

None of the channels will carry advertising, and all the material will be shown without interruptions. The plan will soon be introduced on the cable system in Winter Haven, Fla., and is now available to other systems.

Subscribers to a cable system who wish to receive one or

more of the Gridtronics channels, are given a small converter to place atop their sets, with a four-position pointer to choose the channel they want to receive. They would pay, in addition to their monthly cable rental, $2.50 per month for the Information Channel, $2.50 for the Instructional Channel, $5 for the movie channel, and $2.50 for the medical information channel.

It would be unfair to comment on the content of the Gridtronics package in advance, but nothing in the promotional flier indicates that the offerings will be significantly different in content or approach from material now being produced—with larger staffs, budgets and experience—by existing commercial, educational and public television facilities. And it does little to promote cable TV's greatest potential contribution—its ability to cater to specialized community, minority and individual needs and tastes. In fact, the easy availability of a wide selection of such nationally prepackaged material could inhibit creative local programming.

One of the four channels in the Gridtronics package—that for doctors—*does* carry specialized content, but even here the dangers of commercial packaging are apparent. The channel, says the firm, will "be subsidized by leading pharmaceutical companies." Material will include a "weekly presentation of news" by pharmaceutical manufacturers; a "drug detailing hour" will present "products and their properties by various individual pharmaceutical manufacturers." The drug industry has been waging promotional war on the nation's doctors for years. Now, instead of being visited in person by the industry's detail men, doctors will be solicited by remote control, and will pay for it themselves by the month.

This ambitious venture makes clear that if the nation does not move quickly to formulate comprehensive policies, the shape of cable TV will be largely determined by the forces of the market place.

Another obvious issue raised by the Gridtronics package is how CATV programming is to be financed. A cable system can absorb the costs of local originations on a modest scale, but someone will have to pay for full exploitation of the potential. Edward Greenberg, associate professor of economics at Washington University, St. Louis, discussed the problem in a speech entitled "On the Economic Feasibility of Offering New Services via Cable TV Systems," at the 1969 NCTA Convention. "To attract new subscribers, some programming might be offered at no additional charge. For special events or programming which appeals to fairly small, specialized audiences, additional charges might be made. Finally, advertisers might be willing to sponsor CATV-originated programs."

Professor Greenberg's second item, "additional charges," is actually that recurring nightmare of commercial broadcasting and theatre owners—pay TV. The term is usually not applied to the monthly cable subscription fees, although this in a sense does represent payment for television. "Pay TV," whether over-the-air TV, or by cable, refers to charges made for specific programs. One might well wish that additional payment for cable programming had been restricted, in Professor Greenberg's words, to "special events or programming which appeals to fairly small, specialized audiences," and which would not have reached the TV screen in mass-programming contexts. As it turns out, Gridtronics is initiating pay TV for centrally produced mass programming, in the approximate image of things that are already being done by major broadcasters and educational TV. But the point is not whether this use of pay TV is the best that could have been conceived. The point is that the event so feared by commercial broadcasters and the motion picture theatre industry has in fact occurred.

Because Gridtronics has moved into the mass-programming business, which implies ruffling as few feathers as possible, the firm's position on the movie issue thus far has been to cooperate fully in the existing trade restraint. Stern has agreed that Gridtronics will show on its movie channel only films that have completed their engagements in first-run and neighborhood houses. But there is no guarantee that other cable promoters will be so accommodating. And an even greater danger looms.

As cable systems proliferate, it will become simple to hook them together into a national cable network. Frederick W. Ford, retiring president of the National Cable Television Association, forecast that development in an address to the 1969 NCTA convention. "I expect substantial development very soon in the nationwide distribution of Public Broadcasting Act programs, many of which may be distributed on cable television channels in prime time by way of domestic satellites, which will no doubt interconnect cable systems for nonentertainment type programming."

It will be noted that Ford excluded entertainment programming from his vision of a cable network, but that was because he and his cable association were at the time playing footsie with the National Association of Broadcasters, in which NAB agreed to drop its hostility to program origination by CATV, if the cable men would agree not to hook up their systems to provide mass entertainment. The fact is that there is no reason, other than collusion in restraint of trade, why a cable network should not bring diversity to entertainment programming, as well as other types of programming. And there is no reason, other than collusion in restraint of trade, why such programming should not include first-run movies.

A cable network could operate by assessing each CATV member a small portion of its gross receipts. It could also put out its own special offerings on pay channels. Either way, or in both ways, a cable network would make money, and would be able to move in circles where money talks. It could, among other things, offer fees for first-run films that would tempt at least some movie producers. In addition, it could acquire, for more modest fees, first-run rights to special interest films, to films that have attracted interest and attention at film festivals, and to yet other films whose theatre possibilities would be limited to a few university towns and major cities. All these types of films could earn their way on pay channels, the lower "take" on special-taste films being offset by the lower cost of obtaining the rights.

A cable network with a substantial programming emphasis on films could, therefore, bring to every town and community a variety of fare that is now available only to assiduous moviegoers in a few of the nation's largest cities. Persons could see this material for far less than box-office prices. And this immense new outlet could provide yet another stimulus to diversity and creativity in the making of movies.

This possible chain of developments is what the movie theatres are determined to prevent. Their ingenious and ostensibly altruistic ploy is a campaign to "Save Free TV!"—that is, to save the type of TV that does not bring into the home films in greater variety and at less cost than is possible under the present system of theatre distribution.

The National Association of Theatre Owners (NATO) and its affiliated theatres have mounted an immense and costly effort. Printed material for the campaign includes an "invoice," addressed to "Mr. and Mrs. John Public" and showing a one-month "Pay-TV Bill" of $92.40. From coast to coast, "Save Free TV!" blazes from marquees, and moviegoers are urged to sign petitions in the lobby. Some ten million signatures have been collected, mostly, one can hardly help concluding, from persons who had no idea of what was really at stake.

The industry has also been busy in Washington. Bills to prohibit pay TV have been introduced by twenty-two Congressmen, and an unknown number of bills have appeared in state legislatures and city councils. In New York State, twenty-seven Assemblymen, led by Leonard P. Stavisky of Queens, introduced a bill in February 1969 to prohibit CATV systems from originating movies or other entertainment programs. Hearings were held on the bill in the fall of 1969, and its fate remains in doubt.

In New York City, movie theatre interests pressured the city council into amending the franchises of the city's CATV systems to prohibit them from showing first-run movies. This requirement is now enforced by the city's Bureau of Franchises. The City Club of New York, one of the few civic groups that paid attention to what was happening, expressed its opinion succinctly: "Press freedoms and fair competition are ignored in order to defend old institutions from new technology."

* * *

In addition to payment for specific programs, Professor Greenberg mentioned advertising as a source of funds for CATV programming. It has already come to cable TV—about 10 percent of cable systems accept advertising on their program origination channel.

In at least one sense, that can be regarded as a benefit. Few small merchants or local businessmen can afford over-the-air TV advertising, a circumstance that favors big merchants and the national giants. Cable advertising is much cheaper, and the advertiser, like the candidate for office, can reach his exact local community. He gets more for less.

The real problem with advertising on the cable goes to the heart of what is wrong with the way that CATV is being developed. If a cable operator sells advertising on a channel where he presents his own material, he will want as much of his audience as possible to watch that channel. His motivation will therefore run in the same direction as that of present-day over-the-air broadcasters—that is, it will be against his economic interests to provide as many channels as possible on his cable system, since each additional channel will further reduce the audience for the advertising he has sold.

What is obviously needed is a system that will give the cable operator an exactly opposite motivation—an economic reason to open on his cable as many channels as possible. This involves fundamental issues of public policy, toward which my discussion is directed.

Another current issue is copyright. Cable systems pick up and transmit the entire offerings of over-the-air broadcasts, including copyrighted material. For years the broadcasters have been protesting that the cable system should pay for such material, but the cable owners maintain that, like any other citizen, they are entitled to put up an antenna and receive signals without charge. Broadcasters make the obvious retort that other citizens do not rebroadcast the signals for profit, but so far no copyright fees have been paid.

The copyright dispute has become a pawn in the larger struggle between cable companies and the broadcasting industry. By charging big royalties, by requiring that the cable companies make the payments individually to each of the thousands of persons to whom they are due, and by insisting that back payments be made for all material thus far used, the broadcasters could seriously cripple the cable industry. For their part, the cable operators have made it clear that, if excessive royalties were demanded, they would move more quickly and more extensively into program origination.

Despite their public stance, most cable operators privately concede that reasonable royalties are probably due and must ultimately be paid. This feeling was reinforced in 1966, when the U.S. District Court for the Southern District of New York ruled that two cable systems in West Virginia owned by the

Fortnightly Corporation had infringed copyright by transmitting shows owned by United Artists Television, Inc. When Fortnightly appealed to the Supreme Court, nobody in the cable industry was optimistic about the outcome. Representatives of the cable and broadcasting industries then began private discussions on the copyright problem, with the broadcasters holding most of the aces.

While these discussions were going on, and to almost everyone's surprise, the Supreme Court in June 1968 reversed the District Court's ruling. The key question, the court said, was whether cable systems "performed" the shows, since the 1909 copyright law required payment only for actual performance. A 5-to-1 majority said that retransmission by cable was not a performance.

This victory for the cable interests is only the preliminary round. For some time the Copyright Subcommittee of the Senate Judiciary Committee, headed by Sen. John L. McClellan (D., Ark.) has been working on a general revision of the copyright laws, and for more than a year the committee has been assuring both broadcasters and cable interests that their copyright problems would be dealt with specifically in the new bill.

With that legislation in prospect, and for other reasons that will be evident in a moment, the National Cable Television Association and the National Association of Broadcasters scheduled further private talks to cover not only the copyright problem but a variety of other matters in contention between them. On May 29, 1969, the two groups announced a compromise accord by which cable operators would pay copyright fees and, among other things, would agree not to create a network of cable systems that would originate entertainment programming.

As things worked out, this proposed cartel never went into

effect, and since then it has not been possible to get the parties to agree on a formula for copyright payment.

Two other issues that relate to the place of cable TV in the nation's communications systems are the increased crowding of the electromagnetic spectrum, and the UHF TV channels.

Television is, of course, only one of many users of the air waves. Others include AM and FM radio; shortwave and amateur radio; two-way telephone systems for private cars; walkie-talkies; remote-control systems used by homeowners for such purposes as opening garage doors; short-distance, two-way radio systems for dispatching and directing taxis, delivery trucks and other types of commercial vehicles; the airlines; police and fire departments; the government and the military; marine communications; microwave stations for retransmission of signals; communications satellites; the space program and radio astronomy. The FCC estimates that ten million transmitters are now authorized to use the U.S. air waves. To say that they are jostling one another is an understatement, and the crowding can only get worse.

The heaviest pressure on the spectrum is coming from the group that the FCC calls land-mobile users. These are the short-range, two-way radio setups whereby a dispatcher maintains contact with one or more vehicles. The category includes taxi fleets, delivery and service vehicle fleets, and police and fire departments. (Among police departments, additional pressure is coming from the spread of walkie-talkies. The Commission now licenses two hundred thousand new units in this category annually, and expects that more than seven million of them will be in operation by 1980.

The frequency range best suited for this kind of transmission also includes TV. Within this range, 60 percent of the spectrum is allocated to TV and FM broadcasting, 34 percent is set aside for government and aircraft use, and only 5 percent is allocated to land-mobile communications.

The land-mobile users want the FCC to give them part of the TV spectrum. Their initial requests have been for part of the UHF spectrum. They point out that the UHF band is not being widely used. They state that commercial TV broadcasting is an entertainment and advertising medium, and as such not intrinsically entitled to an overwhelmingly large span of the spectrum at the expense of other legitimate uses. Even if part of the spectrum were reassigned for these purposes, they say, plenty would be left for selling soap.

TV broadcasters vehemently resist the reassignment of any part of the TV spectrum, but there is another possible solution. In centers of population over-the-air television can be transferred to the cable, thus clearing broad areas of the spectrum for land-mobile use.

This solution is no more acceptable to the TV industry than is parting with a share of its spectrum, since it subjects their programming to all the hazards of increased volume and variety of TV fare already noted. It is a plain fact, however, that TV can be transmitted by cable and land-mobile communications cannot. Relieving congestion on the air is another important potential benefit of the cable.

The attempt of land-mobile users to invade the UHF band is a comment on what has been happening in this area of the spectrum. Up through the 1950s, TV signals went out over only the twelve channels of the VHF spectrum which the FCC had assigned to commercial users. The result of this assignment was an inadequate choice of stations for most viewers, near uniformity of fare, and little community service. The pattern was contrary to the public interest, and the FCC could not ignore it indefinitely.

Legally, the Commission could have reassigned part of the VHF spectrum to noncommercial use, but that move was a political impossibility. The FCC therefore took what seemed the next best step: it opened up the UHF spectrum for commercial licensing. The Commission hoped especially that the new UHF stations would move into the area of community service. With strong FCC urging, Congress in 1962 passed a law requiring all new TV receivers to have UHF tuners.

Far from resisting this approach to a broadening of available TV fare, the VHF broadcasters applauded it. Assuming that public policy could not indefinitely ignore the results of the VHF giveaway, this "reform" promised existing broadcasters the least trouble. It provided the illusion that something effective was being done, and could keep the FCC busy for years building up UHF broadcasting. For the VHF broadcasters these would be additional years of grace.

The ultra-high frequencies are poorer signal carriers than the very high frequencies. To produce a signal of given strength and clarity, a UHF station must install more expensive equipment than is required by a VHF station. Few UHF outlets have been able to make this additional investment, and their signals are therefore inferior.

At the same time, UHF stations get less return on the investment they do make, because the VHF outlets are already squarely camped on most existing sources of revenue. UHF, therefore, is always on the verge of starving to death, and the FCC has in fact devoted much time trying to prevent it from foundering. In 1967, 133 UHF stations were in operation, but only forty-four showed profits, and of these, forty-two had network affiliations. Thus, of eighty-nine non-network UHFs, only two were in the black. In Chicago, one of the nation's largest advertising markets, four UHF stations have died.

With little to invest in programming, and with almost no chance to succeed unless they are affiliated with a network, UHFs cannot provide a genuine alternative to existing types of programming. And they certainly cannot mount comprehensive and creative community services of the kind the FCC so naively foresaw. "Is it any wonder," says Washington TV consultant Arthur Alpert, "that the typical UHF station consists of a transmitter, a film projector, old situation comedies and ancient movies?"

But the more incompetent UHF has proved itself as an alternative to VHF, the harder the FCC has pushed it. Behind this unswerving devotion lies an intensive lobbying effort—not by the UHF stations, which have no money to hire lobbyists but by the VHF industry. For them, this approach to better U.S. television is a dream come true.

Americans are so accustomed to their system of broadcasting that it probably occurs to only a few that different arrangements exist in every other country in the world. And if they do notice the difference they assume, no doubt, that our system is the result of clear national choice. Quite the opposite is the case. "Perhaps never before in history," says Harry J. Skornia, professor of communications at Northwestern University and author of *Television and Society,* "have the most powerful channels to the people been so completely controlled by so small a segment of the national life." The point is elaborated by Alexander Kendrick in *Prime Time,* his biography of Edward R. Murrow: "Far from being an expression of majority desire, as the networks say, television programs are the imposition of a social minority on the majority, the minority consisting of the fifty top advertisers, the three networks and a dozen advertising agencies. It is what they think public taste is and demands that governs the nature of broadcasting."

Only in America is this true, because only in America is television basically a commercial enterprise. Back in the years 1948–51, when the FCC was allocating the twelve channels of the VHF, the question arose as to who should get the licenses. At least some staff members of the Commission took the question seriously. A number of major interests in American society might reasonably have claimed a share of the scarce spectrum. Corporations and commercial interests were one;

education was another; nonprofit public programming, subsidized either by the government or by a combination of government funds and private philanthropy, was a third. Labor was certainly entitled to consideration, particularly if part of the spectrum was to be licensed to corporations. Religious groups, and other important interests in society, might merit consideration. But, as it turned out, the group with the real power in Congress and at the FCC—the existing commercial radio industry and radio networks—walked away with the whole pie. Incredibly, the entire VHF spectrum was given to a very small group of immensely wealthy persons and corporations, to be used for the sale of cigarettes, soap and a few other consumer products noted principally for the large amount of advertising expense represented in the retail price.

Educational TV and public TV have belatedly edged onto the fringes of the spectrum, with budgets that depend on the largess of foundations and the uncertain annual whim of Congress, and that in any case would hardly keep U.S. commercial television on the air for more than a few hours a year. The solid sweep of the VHF spectrum is still in the hands of the hucksters, and they will not give it up.

Television is one of the most remunerative activities in the annals of business. Broadcasters average a 90 to 100 percent return on their tangible investment annually. However, some members of this money-making fraternity do better than others. The gold mines are the stations affiliated with one of the three major networks. They hook themselves onto the network line and receive an unending flow of mass-audience material. The network pays them for the audience they contribute, and they can also sell commercials on spots adjacent to network shows. Stations with this arrangement find it impossible not to amass huge profits; except for a few independents in the largest cities, stations without network affiliation usually cannot make the grade at all.

Thus the way to succeed in commercial U.S. television is to get a network contract, do little or nothing to build up local programming, and confine one's efforts to making regular trips to the bank. One need not live in, or be interested in, the community where the station is located. One can buy a string of these little wonders, here and there throughout the country, and rake in the proceeds. That is what is done; that is the shape and the form of television in the United States.

It has had at least two notable effects. First, and catastrophically serious, a nation beset by multiple crises had been deprived of its most potent medium of communication, a medium it could have used in the service of its overwhelming social needs. Second, the content and emphasis of the programming designed by the powerful few to deliver the public to the advertiser may have made at least some of the social problems worse. On the one hand, there has inevitably been, as Skornia puts it, an "authoritarian imposition of attitudes and values manufactured by the business community." On the other hand, in the words of a recent statement by the National Citizens Committee for Broadcasting: "The great majority of broadcast programs are devoted to vulgarity and violence. . . . We believe this inferior programming on commercial TV and radio undermines our society. . . ."

Capturing the VHF spectrum for private exploitation was an impressive achievement by perhaps the most powerful of Washington lobbies. As I said earlier, the broadcasting industry controls an access to the public that is essential to all members of the government. In addition, at least thirty Senators and Congressmen are owners or part owners of TV or radio stations, and a far larger number have interests either in corporations that own or operate broadcast properties, or in closely allied communications industries. Broadcasters have long had their way in matters of legislation. Says Skornia: "The fact that in recent years so large a proportion of legislation has been devoted to protection of property and of the rights of broadcasters, rather than to protection of the consumers, reveals the power of the broadcast industry to define regulation and effect legislation on their own terms."

A current instance is the FCC's activity on the license renewals of certain TV stations. Recently, civic groups have been formed in several communities to bid for the licenses of certain TV stations when they came up for renewal, the interest being to turn the stations into better vehicles of community service. The law requires the FCC to hold hearings whenever such a challenger comes along, and it has held a number of such hearings. In one 1968 action, the Commission refused to renew the license of station WHDH-TV, owned by the *Boston Herald-Traveler*. The principal reason given was undue concentration of media ownership.

But cross-ownership of media is more the rule than the exception, and the frightened broadcasters turned to Congress for help. Congress responded quickly. Sen. John O. Pastore (D., R.I.), chairman of the Senate Commerce Committee's Subcommittee on Communications, introduced a bill to reverse the requirements of the present law. It would prohibit the FCC from holding hearings when a station whose license was up for renewal was faced with a challenge. Instead, the FCC could consider a new application only after it had cancelled a license. This would prevent the FCC from comparing an existing station's programming record with proposals submitted by competitors for the channel.

The FCC got the message. On January 15, it issued a statement reassuring broadcasters that it would not consider license challenges against radio or TV stations that "substantially" met the programming needs of their communities. Challenges to such stations would be dismissed without reference to other issues, and their licenses would be renewed. "Other issues" obviously include cross-ownership of media, and a challenger's promise of better programming. The statement, of which Dean Burch, Nixon-appointed FCC head, was the principal author, was endorsed by a 6–to–1 majority. Commissioner Johnson was the only dissenter.

The NAB was jubilant, but served notice that it also wanted the even more stringent Pastore bill. "The policy statement is great," said Vincent T. Wasilewski, the Association's president, "but we still feel that there is a necessity for the legislation."

The broadcasters' jubilation, however, was short-lived. The FCC ruling was challenged in court by Citizens Communications Center, one of Ralph Nader's consumer protection groups, and by Black Efforts for Soul in Television, a watchdog of minority interests in TV. On June 11, 1971 the Court of Appeals in Washington, D.C. overturned the FCC rule, stating that it "violates the Federal Communications Act of 1934, as interpreted by both the Supreme Court and this court." The court cited "explicit language in the Act to the effect that no automatic preferential rights are intended to be extended to the renewal applicant," and the Act's requirements for a full hearing to competing applications.

The broadcasters now turned to the White House, and again their voice was heard and obeyed. In a major policy speech to the International Radio and Television Society in October 1971, Clay T. Whitehead, director of the White House Office of Telecommunications Policy, proposed that legislation be enacted that would extend the duration of TV licenses beyond the present three-year period, and would permit FCC to entertain competing applications for a chanel only when a license was revoked or not renewed.

In this fashion Congress and the Executive branch of the Government, responding to the broadcasting lobby, keeps the FCC in line. It also plays the broadcasters' game by keeping the Commission small and underfinanced.

Before he assumed office, President Kennedy commissioned

a number of studies and task force reports. One on the FCC, prepared by James Landis, stated that no other governnment agency was subjected to pressures of the intensity of those exerted on the FCC; that no other agency bowed more completely to the pressures; and that no other agency was in greater need of being freed from them. More than ten years later, the need is greater than ever.

4 *The Cable and the Regulators*

Cable television men enjoy depicting the FCC as an ogre intent on destroying their industry. The source of the Commission's animus is alternately ascribed to an excessive lust for power and to its role as the humble servant of the broadcasters and other powerful segments of the communications industry. Describing recent FCC policies relating to cable TV at the 1969 NCTA Convention, general counsel Bruce Lovett told the audience that, by its actions, "the Commission demonstrated its intent to constantly probe our industry for new thresholds of pain and to continue a policy of oblivion for cable television."

The picture is overdrawn. It assumes a higher degree of malevolent intent, and a greater capacity for carrying it out, than the alleged villain has ever possessed. Such a theory cannot account for the complete fragmentation of opinion among the commissioners, revealed by separate statements attached by a number of them to the Commission's basic regulatory orders, and to decisions in a number of cases arising under these orders. Neither can it account for certain individual actions of the Commission. For example, when TelePrompTer Cable TV sought FCC permission to experiment with transmitting cable signals for short distances by over-the-air microwave relay in crowded sections of New York City, where the laying of cable is very costly, the FCC granted the request over vigorous objections by AT&T, Comsat and Western Union, all of which want to develop this area of technology themselves, and all of which have plenty of friends in Washington.

Nevertheless, when every allowance has been made, the FCC's activity in the field of cable TV, at least until 1970, must be regarded as a serious failure. The problem stems principally from the fact that the Commission has defined the public interest as the perpetuation of the over-the-air television industry as it now exists. "I think you should remember the fundamental hypothesis of the Second Report and Order adopting our CATV rules in March of 1966," Comr. Kenneth A. Cox said to an NAB group in 1968. "It concluded that rules were necessary in order to keep CATV a supplemental service, rather than risk its displacing our basic over-the-air system." FCC's goal, Commission chairman Rosel H. Hyde told members of the House in 1969, has been "to integrate the CATV operation into the national television structure in a manner which does not undermine the television broadcast service." In this context, anything that threatens to "undermine the television broadcast service" becomes "unfair competition."

The shortcomings of this approach are obvious. First, the assumption that the public interest lies in preserving the present setup is open to challenge. Many knowledgeable observers hold an almost opposite view. Second, by using this concept as a basis for making policy, the Commission becomes a party to economic struggles between competing interests and technologies. "Protecting existing interests from the superiority of a new form is a derogation of the Commission's statutory responsibility," said the American Civil Liberties Union in a brief on cable TV filed with the Commission in 1969.

Third, this approach assumes that the public interest is divorced from what the public wants. Indeed, it assumes that the more the public wants the cable, the more vigorously the Commission must act to prevent the public from having it, since high public demand increases the likelihood that the cable will threaten existing broadcasting interests. Last, and far from least, preoccupation with "unfair competition," has involved the FCC in control of what a cable system may carry. Such action raises fundamental issues of free speech and freedom of the press, and it is puzzling that this aspect of the FCC's cable policy did not result in a barrage of court challenges years ago.

When cable systems first appeared, broadcasters welcomed them because they increased the potential audience, but it soon became evident that the effect would be more complex. By bringing a greater number of channels and a greater diversity of fare to a given area, CATV fragmented the audiences that local stations regarded as their private property.

The first agitation by broadcasters to control cable TV came during 1958 Senate Commerce Committee hearings on the problems of bringing TV service to smaller communities. Twelve TV station owners who were called to testify begged Congress to protect them from the cable systems that had invaded their bailiwicks.

This familiar cry of alarm has never been supported by real evidence. For example, none of the twelve stations that in 1958 asked Congress for protection has gone out of business. Despite the cable, all have fared well, and have substantially increased their advertising rates. In the cities in which two of the stations are located, additional TV stations have gone on the air. In two other cities, the owners who subjected Congress to such a woeful picture of their predicament, have since sold their stations at substantial profit.

In the larger picture, cable TV has quadrupled since 1958,

both in number of systems and in audience size. In addition, most cable systems, which offered only five channels in 1958, can now carry twelve, a flexibility that further increases the "competition." Nevertheless, 173 new commercial and 120 educational stations have gone on the air in this period, 44 of them in communities where cable systems were already operating.

Having made its pitch to Congress, the broadcast industry pounded on the doors of the FCC. In 1959 the Commission ruled that it did not have jurisdiction over cable TV. Congressional reaction was prompt. Senator Pastore introduced a bill providing for FCC control of cable TV through licensing procedures. The bill (S.2653) narrowly failed, but FCC got the point. It was obvious, the Commission realized on re-examining the problem, that it had jurisdiction at least over cable systems that used over-the-air microwave relays.

In 1962 the Commission began to rule on individual cases of cable systems using such relays, and its authority to do so was upheld by the courts. In 1965 the Commission issued its "First Report and Order," setting forth certain general rules for microwave-served cable systems. In 1966 it issued a "Second Report and Order," asserting its authority over all types of cable systems, whether or not they were served by microwave relays. The Second Order applied the rules of the First Order to all cable systems, and added some new rules. Its general assertion of authority over all cable TV, and the specific rules contained in its two orders, were again sustained by the courts.

In 1968 the Commission issued a "Notice of Inquiry and Proposed Rulemaking," setting forth some proposed new rules, listing a number of questions about the growth and direction of cable, and soliciting comments on both from the communications industry and from social scientists. Opinions, most quite strong, poured in.

In July 1970, the Commission issued a "Second Further

Notice of Proposed Rulemaking," adding new proposals to its 1968 Notice, and also started four more rulemaking proceedings relating to cable. In the Spring of 1971 FCC held five days of televised hearings on the Second Further Notice and the new dockets. It then began to draft new rules. But by then two additional parties had entered the struggle to control the destinies of the Wired Nation—Congress and the White House.

Each of FCC's major rulemaking efforts—The First Report and Order, the Second Report and Order and the Notice of Inquiry—is encompassed in a long, sometimes confusing, document. FCC policy is further embodied in decisions in a number of cases and controversies. The major requirements that FCC has placed on the cable can, however, be set forth without entering too far into the jungle of subsidiary issues.

The First Report and Order. This set forth what are usually referred to as the carriage and non-duplication rules. The carriage rule requires a cable system in a given community to carry the local TV station, along with whatever other stations it may offer. The purpose is to insure that cable subscribers are not cut off from their local TV outlet. (A TV set hooked to the cable can no longer receive over-the-air signals.)

The nonduplication rule concerns the importation, from a distant station, of a program that is being shown by the local station. Such action, FCC Chairman Hyde said in his 1969 testimony, "fractionizes the local audience, and audience is what the local station in effect sells. . . ." The rule prohibits a cable system from showing such programs on the same day that the local station shows them.

The carriage rule has aroused little controversy. There are reasons why a television viewer should not be deprived of local broadcasts when he goes on the cable, and in practice few cable systems ever intended to do this to him. The nonduplication rule is now also accepted with little grumbling, but it launches the Commission much further into the fields of economic protectionism and control of TV content. The supposition on which it rests, that local stations may die for lack of sufficient advertising revenues if certain of their offerings are also delivered to cable viewers from a distant station, is entirely undemonstrated. But even if such drastic consequences could be proved, efforts to prevent it by controlling the content of the cable raise issues at least as serious as those that the policy is designed to solve.

Adoption of the First Report and Order by the Commission was not unanimous; Commissioners Robert T. Bartley and Lee Loevinger issued dissenting statements. Loevinger's was eloquent, and went to the heart of the issue. "The Commission is doing the wrong thing when it seeks to control, directly or indirectly, the specific programs which shall be presented to the audience. . . . I disagree with the nonduplication rule which I believe is an improper attempt to limit competition by controlling programming. . . . Apparently the Commission has not yet learned that expansion of service is not to be attained by the limitation of competition and the imposition of rigorous regulation, but rather by stimulating competition and moderating regulation."

The Second Report and Order. The FCC nevertheless went much further down the path of program control in its Second Report and Order in 1966. The most important provision of this document relates to cable systems that propose to operate in any of the nation's one hundred largest urban market areas. The rule states that if any established TV station in the market area, or even any group that *intends* to establish a new UHF station, objects to the cable system's plan to import distant signals, then the cable system cannot import them unless it can prove to the Commission's satisfaction that such importation "would be consistent with the establishment and healthy maintenance of TV broadcast service within the area." Existing or projected stations in the area do not need to offer any evidence that importation of distant signals will harm them; they need only object. The cable system must then prove that such importation will do no harm to established local interests. Such proof, obviously, is not easy to get.

The practical effect has been to prevent most cable systems in the nation's top one hundred market areas from importing distant signals. This, in turn, deprives cable TV of its principal selling point, and has substantially curbed the entry of the cable into most of the nation's urban areas. The one hundred top TV markets comprise about 90 percent of the nation's population.

The adoption of the Second Report and Order further fragmented the Commission. Of the seven Commissioners, Robert T. Bartley, Kenneth A. Cox and Lee Loevinger, issued statements dissenting in whole or in part. Loevinger was at his colorful best, attacking not only the content of the order but the faulty procedures by which the majority arrived at its conclusions, and the way in which the conclusions were expressed: "The opinion refers to little evidence and much of that is inconsistent; assumptions and speculation substitute for facts, the reasoning is circuitous and illogical, and the statement is so turgid and prolix that it does more to obscure than illuminate the subject."

An interesting feature of the Second Report and Order is an apparent shift in emphasis. In the 1958 Senate hearings and in the First Report and Order, the focus was on protecting small local stations. The politically important pressure, however, was coming from the big stations in major urban areas. For a time the pretense of being concerned for the well-being of smaller stations was a useful way for the major powers in broadcasting to curb the growth of cable, but it eventually became clear that action to protect major broadcasters in urban areas could not be based on an assumption of altruism. The smaller stations were therefore cut off and left to their fate. Under the Second Report and Order, cable systems are free to import distant signals into communities not within the one hundred top markets.

In the Second Report and Order, the FCC also returns to its favorite project, fostering the growth of UHF. "Independent UHF stations are coming on the air in many of the larger markets," FCC Chairman Hyde explained in his 1969 testimony, and they need to be protected from the competition of the cable. Thus a group that even alleges an intention to open a UHF station in one of the one hundred largest markets can prevent a cable system from importing distant signals.

There is, however, little evidence that this will nourish UHF. In fact, the exact opposite is probably the case. Frazier, Gross and Co., a management consulting firm in Washington, D.C., has found that CATV homes watch more TV than non-CATV homes. Because of this, and perhaps also because UHF enjoys equal clarity and equal position on the cable dial with VHF stations, "CATV located in a mixed market and importing distant signals can have a salutory effect on the audiences of the local UHF station[s]."

The growth of UHF is impeded by tremendous handicaps already mentioned and it will probably never prosper no matter what cable systems or the FCC do or fail to do. The true beneficiaries of the Second Report and Order are not the UHF outlets but the VHF giants.

The Second Report and Order has created a nationwide patchwork of absurdities. A typical case is the cable system in Greensburg, Pa. Greensburg lies within the Pittsburgh market area as defined by FCC, and therefore its cable cannot offer distant signals to its subscribers if Pittsburgh broadcasters object. The system's request to do so was blocked by Pittsburgh's KDKA-TV, one of the nation's biggest stations, and by Springfield Broadcasting Company, which has a permit to build a UHF outlet in Pittsburgh.

The Greensburg cable carries all Pittsburgh stations, plus one station each from Steubenville, Ohio and Johnstown, Pa., which are regarded as being within the Pittsburgh market area. It would like to import two other stations, one from Wheeling, W. Va., and one from Altoona, Pa. These stations can be received, with varying degrees of clarity by many residents of Greensburg, but they are kept off the cable, and this has substantially reduced the number of subscribers.

The big loser in this situation, apart from the cable system, is Channel 53 in Pittsburgh, a UHF outlet. Channel 53 is happy to be on the Greensburg cable, where it can compete with the VHFs, but the cable, because it is unable to deliver the Wheeling and Altoona signals, cannot provide Channel 53 with as big an audience as it otherwise might. The winner, of course, is monster KDKA-TV.

In San Diego, the FCC "protected" the local UHF outlet by prohibiting the cable from carrying Los Angeles signals. The cable operator put up a lively fight, alleging that the FCC had no jurisdiction over CATV in the first place. The case, *Southwestern Cable Corp.* v. *FCC,* went to the Supreme Court, which upheld the Commission's jurisdiction. Interestingly enough, the fight in San Diego was not waged by the UHF station but by the local VHF station. The FCC found that if the cable carried Los Angeles signals, it might be able to sign up as many as half of all San Diego set owners. This would represent "substantial penetration" of the San Diego market which, in turn, made FCC prohibitory action the more urgent.

In San Diego, the Commission also prohibited the cable system from carrying advertising on its program origination channel. In some cases, the FCC has prohibited such advertising, and in others has permitted it. The criterion has been the Commission's opinion as to whether or not the cable's acceptance of advertising would cut significantly into the revenue of over-the-air local stations. Again the Commission had little evidence to go on. And again it has found itself in the position of arbitrating between competing interests.

The Proposed New Rules and the Notice of Inquiry. FCC's proposed new CATV rules and its list of questions about the role of cable in the nation's communications future, both promulgated on December 13, 1968 (with additions and clarifications on January 17 and May 16, 1969), represent a genuine effort by the Commission to rethink its role in cable TV. The American Civil Liberties Union, one of many organizations that took advantage of the invitation to file opinions, stated that the results of this rulemaking procedure "will surely affect the American system of public communications for decades to come."

On the importation of distant signals, the proposed rules, which were enacted on an interim basis by the Commission pending the completion of the proceeding, are even more restrictive than the provisions of the Second Report and Order. Under the Second Report and Order, a cable system that thought it could convince the Commission that such importation into a major market area would not hurt local broadcasters could at least ask for a hearing. Under the new rules, importation is prohibited in the top one hundred markets unless the operator has obtained, on a program-by-program basis, the original broadcaster's consent to retransmission. Even if the original broadcaster were inclined to give this consent, his contracts with the persons who hold the copyrights would prevent him from doing so. For all practical purposes, therefore, the new rule reimposes on cable systems in major market areas the copyright liability that was struck down by the Supreme Court in the *Fortnightly* case.

Actually, there is good reason to believe that this section of the rules represents an effort to plug the hole in the copyright dike that was opened by the *Fortnightly* ruling, until such time as Congress passes new copyright legislation. To one dubious role, that of protecting the nation's broadcasters, the FCC added a second and even more farfetched one—that of protecting the nation's copyright holders. And in this case it is imposing by administrative indirection a requirement whose direct enforcement was ruled illegal by the Supreme Court.

Fortunately, the rules are better and more interesting in other areas. One of these is program origination. After vacillating for a number of years, FCC came out solidly for it despite opposition from many segments of the broadcasting industry. In fact, one provision of the rules would have *required* all cable systems with more than 3,500 subscribers to originate programming. The cable industry has gotten a lot of public relations mileage out of the cable's potential for local and community programming, but when FCC proposed to require such programming the industry fought back vigorously. Nearly half of the systems affected by the requirement asked to be excused from it, and one, Midwest Video Company, challenged the rule in court.

In May 1971 the U.S. Court of Appeals ruled that, however constructive FCC's intent may have been, the rule exceeded the Commission's authority under the Communications Act and could not be promulgated. At this writing FCC has the ruling under appeal, perhaps more to test the limits of its jurisdiction than to enforce the rule rigorously if it should win. Actually, both within and outside the Commission there are an increasing number of persons who think that the best way to realize cable's potential may be to phase cable operators out of programming instead of forcing them to get in.

But if the cable operator doesn't do the programming, who will? The hint of an answer may lie in another provision of the proposed rules, under which the operator would be "encouraged" to make any empty channels on his system available on lease to all comers, on a "common carrier" basis. Here, no bigger than a man's hand and tucked away in a few lines of a long and complex document, is probably the most momentous issue of all. "If they had *required* common-carrier status and made program origination optional, instead of the other way around," says Sidney W. Dean, Jr., a specialist in marketing and communications and chairman of the Communications Committee of the Americans for Democratic Action, "then we would be in the right ball park at last." We shall have more to say about Mr. Dean's ball park later on.

In the proposed rules the FCC also took its first steps into another important area—ownership of cable companies. Belatedly, and on a small scale, the Commission showed signs of recognizing what Commissioner Johnson observed in 1968: "Technological change can bring change in structure, but it is as likely to be change to even greater concentration as to wider diversity."

One big owner of cable companies is the broadcasting industry. Faced with a medium that could offer serious competition, they have been buying the cable systems, or building them from the start. At present about a third of U.S. cable systems are owned by over-the-air broadcasters. In many instances the ownership patterns involved stations and cable systems in the same community. A typical holding was that controlled by the family of former President Lyndon B. Johnson. Ladybird, Linda, and Luci owned 84 percent of the stock of Capital Cable Company in Austin. Mrs. Johnson also owns Texas Broadcasting Corporation, which holds the licenses for KTBC–AM, KTBC–FM and KTBC–TV in Austin.

During the 1960s the networks also moved in. Owners of cable systems included CBS, NBC, Westinghouse, and Cox Broadcasting Corporation. Other enthusiastic buyers of cable systems are newspapers and magazines. Both are aware that the cable can transmit printed matter directly into the home. Time, Inc. has invested $15 million in cable TV. All in all, half of U.S. cable systems are now owned by other communications media and the percentage is continuing to rise.

Still other buyers are equipment manufacturers such as General Electric, and giant industrial combines such as Kaiser Industries and Gulf & Western. The latter, one of the largest conglomerates, holds cable franchises in twenty-seven cities, in twelve of which it now has systems operating. Gulf & Western also owns Paramount Pictures. Asked if he intended to use Paramount films on the cable systems, G&W Chairman James Shaw replied: "Yes. This is why we think we can offer more to the people of this country than any other MSO [multi-system owner]."

Within the cable industry itself, ownership concentration is also increasing, and absentee ownership becomes more common. Until recently the largest owner was H&B American Corporation, whose principal stockholder, Jack Kent Cooke, owned a 32.5 percent interest in the company. H&B, which went into business in 1960, owned sixty-two cable systems; it also owned a radio station in Santa Barbara County, California, was an applicant for UHF Channel 16 in Dubuque, Iowa, and was half owner of Television Testing Company, which uses CATV for market research for advertisers and broadcasters.

In 1970 this giant was swallowed whole by TelePrompTer Corporation. TelePrompTer now owns 130 systems. It also recently acquired Filmation Associates, Hollywood, a producer of animated cartoons and commercials, which the firm intends to use as a "program source for television and CATV." In addition, TelePrompTer is involved in master-antenna TV and closed-circuit design and installation, and is a major provider of facilities for large-screen telecasts for business meetings, sports and special events. The principal stockholder of TelePrompTer is Hughes Aircraft, with 17 percent of the stock.

As multiple systems have grown, the owners of a number of them have tapped an additional source of capital by going public. They thus create a large new group of persons with a financial interest in cable enterprises.

"When a cable company goes public," says Robert W. Wigod, a senior securities analyst with Paine, Webber, Jackson & Curtis and a specialist in CATV offerings, "it frequently must reorient its thinking to become compatible with the objectives of the broad body of stockholders."

The principal objective of stockholders is "growth." They want the company to increase in size and earnings as quickly as possible. This puts the MSO under pressure to do, not what will be of most value to the viewers of any individual cable system that it owns but what will increase the firm's assets most rapidly. Such growth is far more easily achieved by acquiring new systems, and by introducing costly new services on present systems, than it is by fully developing each system's subscriber list, existing services and individual requirements.

Mr. Wigod expects a rapid acceleration of the pattern. "The likelihood is that within two or three years there will be about ten publicly owned companies, each with more than 200,000 customers.

"Each will have a fairly strong central management with the ability to effect economies of size. . . . At that time it may be a lot harder for the CATV company with 20,000 or 30,000 customers to go public; but, on the other hand, there will be a very attractive and aggressive market for those who want to be acquired by the larger companies." The autonomous, locally owned cable system appears to be going the way of the buffalo.

The FCC's reaction to the situation has been tardy and weak. In April 1965 it issued a Notice of Inquiry on the matter of "acquisition of community antenna television systems by television licensees." In August of that year it followed with a report, stating that the danger of abuse was not yet "sufficiently great to warrant an overall or across-the-board prohibition against cross-ownership of CATV systems and television stations."

In 1967 the FCC began to look into the matter again. The National Association of Broadcasters rushed to the barricades. Congress, it said, must ultimately decide the role of cable TV. Until it does, restrictions on ownership would be premature. "Accordingly," said the NAB, "we respectfully request that no action with regard to promulgation of CATV ownership rules be taken at this time." The FCC soft-pedalled its activity.

In its 1968 proposed rules the FCC finally took a stand. One rule, which it formally promulgated in 1969, prohibited a TV station from owning a CATV system in the station's area of coverage. Another prohibited networks from owning cable systems. These rules were modest enough, but FCC bucked massive opposition to put them into effect. Of more than two hundred parties filing comments on this section of the rule-making proceedings, only two, the Justice Department and Professor Stephen Barnett of the University of California Law School, supported the ownership restrictions. The National Association of Broadcasters and the National Cable Television Association, hard-swinging antagonists on many other issues, were united against these requirements, and there has been serious discussion of legislation to reverse at least the first of the two. It is an issue on which the Commission needs public support which it has so far not received.

Second Further Notice of Proposed Rulemaking and the New Dockets. Toward the end of the 1960s awareness of cable began to increase in the business community, at all levels of government, among foundations, in the academic world, in civic and community organizations, and among the public. This growing awareness was reflected in FCC's new round of rulemaking which it launched in July 1970. On the surface it appeared that FCC was still heavily involved in the juggling of economic interests. Its objective, it said, was "to find a way of opening up cable's potential to serve the public without at the same time undermining the foundation of the existing broadcast structure." In reality, many persons at FCC had become convinced of cable's importance, and were seeking a way to permit cable to grow without arousing the broadcast industry and its allies in Congress to complete intransigence.

After issuing its new proposals and receiving hundreds of new comments and responses, the Commission held five days of open hearings on cable in March 1971. Before a large audience in an auditorium at the National Academy of Sciences, the nine Commissioners sat at one table on the stage, while another table was occupied by successive panels of representatives of the broadcasting and cable industries, technicians, officials of foundations that had sponsored cable studies, academic experts, and leaders of national, regional and local civic groups. Among the many proposals made at the hearings was that of the Ford Foundation, which advocated that at least some urban cable franchises should be granted to local civic, educational, and community groups instead of to profit-making firms.

As usual, the economic struggle centered on the issue of distant signals. Several witnesses wondered why, in the absence of a positive showing of harm done by a cable company to a local TV station, FCC should be concerned with the matter at all; and, where real harm could be shown, why FCC did not provide appropriate relief on a case-to-case basis. Commissioner Johnson asked an even more pointed question. Suppose, he said, worst came to worst and a few marginal TV stations were put out of business by the cable. Why was this considered such a high price to pay for the advent of a new communications medium that offered so many benefits to the public?

Such approaches of course have no chance politically. And in fact, the rulemaking proceeding quickly aroused the interest of Senator Pastore, Chairman of the Senate Communications Subcommittee, and Congressman Torbert Macdonald, Chairman of the House Communications Subcommittee. Both notified FCC Chairman Dean Burch that they did not want the Commission to promulgate any new rules until they and their committees had a chance to review them. Burch acquiesced and on August 5 sent a fifty-five-page letter to both committee chairmen embodying the Commission's proposals. They included the following:

Distant Signals. FCC proposed a complex formula that would, in effect, permit cable systems in the top one hundred markets to import two distant signals, and in a few cases three.

Channels. Urban cable systems would be required to provide a minimum of twenty channels. For every commercial TV signal carried, the systems would have to provide a channel for transmission of other types of material. A public access channel would be required, available without cost to noncommercial users. A channel would also have to be made available, without charge for the first five years, for State and local government use, and a similar channel would have to be made available for educational groups. At the system owner's discretion "excess" channels could be made available on lease to all potential users. Systems would be required to ensure that enough channels are available to meet "all reasonable demand," and some form of two-way communications capacity would be required.

Other Requirements. Local franchising authorities would be required to set "reasonable deadlines" for construction and operation of cable systems after the franchise has been granted. This requirement is intended to put an end to the widespread practice on the part of cable companies of seeking and securing franchises and then doing no building for months or years, until it suits their convenience. In other instances companies have been securing franchises with no intention other than selling them. As to the term of the franchise, FCC suggests a limit of fifteen years. Many are currently granted for much longer terms—"an invitation," FCC says, "to obsolescence." Franchise fees should run between 3 and 5 percent as a maximum; if the fee exceeds 3 percent, the city fathers would be required to show FCC that the fee relates to the actual cost of the local regulatory program.

The proposed rules were promptly denounced by the broadcasters, for whom the importation of even two signals is more competition than they wish to have in the lucrative major markets. Disappointed with the FCC, they took their problem to the White House. President Nixon appointed a cabinet-level committee on cable, headed by Clay T. Whitehead, director of the Office of Telecommunications Policy, to go over the same ground and study the same problems that FCC had studied for a year before issuing its proposed rules. At this point, however, broadcasters and cable interests realized that their only remaining alternatives were to accept a compromise or have the whole matter turned over to Congress for a legislative resolution whose shape and form no one could predict. Shying away from these unknown perils, both sides agreed in October to accept a Whitehead-sponsored plan that follows FCC's distant signals proposals in most respects but gives the broadcasters more power to prevent or postpone the showing on cable of certain programming to which the broadcasters have purchased exclusive rights. The matter was then passed back to FCC for action and promulgation of rules.

The FCC's New Rules. On February 3, 1972, the FCC finally issued its new cable rules, in a document running to some five hundred pages. They contain the provisions set forth above, and incorporate the Whitehead-sponsored industry compromise.

The requirements for non-broadcast channels, for public access channels, and for leased channels, and the various franchising requirements, are valuable and constructive steps toward realizing the potential of this new form of communications, and the FCC deserves praise for its work and its achievement. But in the areas of direct conflict with broadcasting —distant signals and programming—the rules reveal the FCC's continuing inability to free itself from the economic interests of existing commercial TV.

In major markets a small ration of distant signals will be permitted, but in accordance with fantastically complicated formulas that, in practice, add new absurdities to the old. For example, in Washington, D.C., cable systems will be prohibited from carrying Baltimore signals, even though most Washington residents can pick them up with rooftop antennas or even with rabbit ears.* In many smaller cities and towns, which have most need for supplemental TV fare, new systems will be limited to carrying four commercial signals, local or imported. The Federal Government remains firmly committed to preventing cable from overcoming the inequities in the U.S. commercial system of broadcasting.

The provisions of the Whitehead compromise are equally dubious. Broadcasters in the top fifty markets can buy exclusive rights to programming for many years or forever, regardless of how often the broadcaster actually uses the programming, or, indeed, whether he ever uses it at all. When a broadcaster has bought such rights, neither a cable system nor anyone else can secure rights to show the material in the geographical area covered by the broadcaster's signal. Neither can a cable system import the programming as a distant signal from another city,

* This at least offers the hope that cable will finally be freed from regulatory restraints on a future day when some Congressman living in Washington subscribes to the cable and discovers that, because he is hooked up to this great electronic marvel, he can no longer watch the Baltimore Colts!

into a city where a local broadcaster has sewed up exclusive rights.

These requirements have two purposes: to support the high price structure that now pertains for the sale of exclusive rights to programming, as desired by owners of programming material; and to make cable systems in large markets less attractive to subscribers by limiting the variety and diversity of what they can show, as desired by broadcasters. "This serves no one's interests," said Commissioner Johnson in dissenting from these portions of the rules, "save the most powerful broadcasters and program owners, who now get their way."

The rules do not prohibit the building of cable systems in the nation's top fifty markets, but they are designed to retard it, and there is little doubt that they will succeed in doing so. It is in these areas that cable can perhaps bring the greatest social benefits to the nation. But it is also in these areas that the broadcasting and programming industries, as presently constituted, make their biggest killings—syndicated program suppliers, for example, make 80 percent of their total income in the top fifty markets, much of it through the sale of exclusive rights, which cost nearly twice as much as the right to show a program on a non-exclusive basis. Because broadcasters and programmers see cable as a threat to their profits in the cities, the FCC cannot give a green light to the Wired Nation.

Actually, wiring up major cities will be costly, and Federal incentives, rather than disincentives, will be required if the job is to be done. But on this issue Congress seems no more able to act than the FCC.

Before dealing with these long-range policy questions, we must first review the activities of one more major interest group—the nation's telephone companies—and then glance at current action on the state and local levels.

5 *Mysterious Ma Bell*

American Telephone and Telegraph owns outright sixteen telephone companies and is part owner of eight others. Collectively, they are known as the Bell System and they provide 85 percent of the telephone service in the United States. The remaining 15 percent is supplied by 2,500 independent companies that tie into the Bell System for service outside their areas. Bell has 872,000 employees and assets of $40 billion, three times those of General Motors and more than the governmental assets or the annual gross national product of most nations. AT&T is regulated by the Common Carrier Bureau of the FCC. This bureau has 160 employees, including clerical staff, and operates on an annual budget of less than $2 million. Its regulatory responsibilities include, besides the Bell System, fifty-four phone companies that operate across state lines, plus all other interstate communications common carriers, such as microwave relay systems, Western Union and Comsat.

The term "common carrier" has already appeared briefly in these pages; it will play a bigger part from now on, and at this point it should be carefully defined. Common-carrier status is assigned to enterprises offering services deemed necessary to the public, and whose nature dictates that they be run as monopolies or near monopolies. When a communications system is designated as a common carrier, two things follow. First, it must be accessible, on a nondiscriminatory basis and at standard rates, to anyone who wishes to use it. Second, the owner or operator may not interfere with the content of what goes over his system. The telephone system is a common carrier; broadcasting and cable TV are not.

For many years the telephone company owned the only communications wire going into the American home. Now a second one, the cable, is poking its way through the walls. And the cable is not just a second wire but one with far greater communications-carrying capacity than the telephone wire. "If a real broadband [coaxial cable] network is ever constructed," says former FCC Commissioner Kenneth Cox, "its operators could virtually provide conventional voice telephone service for nothing."

And much more than conventional phone service is at stake for AT&T. It is already involved in the common-carrier transmission of many types of electronic information other than voice-by-phone. For example, long-distance transmission of network television shows, TWX communications, various kinds of data, and exchanges of information between computers, go over phone company facilities at scheduled rates. Phone companies expect their greatest growth in the next decade to come from such nonvoice types of transmission, and from such new services as Picturephone, rather than from expansion of ordinary telephone service. AT&T also keeps a sharp eye on such developments as the conducting of financial transactions by wire, and the home communications center.

Thus Bell would dearly love to extend its ownership to the second communications wire into the American home. Like those other fearless champions of free enterprise, the broadcast industry and the newspapers, the phone company prefers life without competitors. It reacted to the coming of the cable in the same logical way as have those other two industries: by seeking to own cable systems. AT&T may nevertheless have lost its bid to own the physical plant of the Wired Nation. The struggle has been one of the least known and most important to take place in Washington in recent times.

In working out its approach to cable TV, the telephone industry found itself in an odd position. In 1956 AT&T entered into an antitrust consent decree with the Justice Department, in which the company agreed not to conduct any kind of business other than common-carrier operations. The decree did not apply to the independent companies.

Since CATV systems do not at present have the status of common carriers, Bell System companies are excluded from being the actual operators of cables. Some of the independents have been operating cable facilities, but Bell must move more circumspectly. The device it uses is to build cable systems and let other parties run them on long-term lease. Telephone companies would like to own as many cable systems as possible. That would make them *de facto* owners of substantial portions of the Wired Nation network, if it comes into existence. In time the long-term leases will expire, and by then the FCC and Congress might be willing to let telephone companies own the Wired Nation, in whole, or in part.

A weapon the phone companies use to this end is their ownership of the poles or conduits by which their wires are brought into individual homes. It would be impractical for cable systems to create duplicate systems of poles or conduits, and it is unlikely that any municipality would permit it. Therefore, cable operators must negotiate an agreement with the phone company.

In some instances the phone companies have simply refused to grant pole-attachment rights; more often the situation has been one of interminable delay. Months go by while papers are shuffled. When the phone company finally does produce a schedule of rates, it may turn out to be disagreeably higher than the cable operator had expected.

There are other problems. The phone company, for example, has an easement to bring its lines onto private property such as apartment buildings. The cable operator does not share in this easement, and state governments, which have made easement rights available to phone companies, have given no such help to cable companies. The cable operator who builds his own system enters a jungle in which apartment building owners, or owners of rented homes, may refuse outright to let the cable come onto their premises, or may grant permission only upon payment of what amounts to a substantial bribe.

While fending off the cable operator who wants to build his own system, the phone company hopes that someone will come along who will ask the phone company to build and lease back the rig. Or the operator who originally wanted to build his own system, may throw in the sponge and ask the phone company to build and lease him the cable.

If that happens, all problems, obstacles and delays instantly vanish. The system is built quickly; the cable goes up on the poles; the phone company uses its easement rights to get it to its destinations; and the completed system is promptly turned over to the lessee.

Independent cable operators have not taken lying down the incursion of telephone companies into the cable field. Their most useful weapon to date has been Section 214 of the Communications Act, which states that if a phone company wishes to offer a given type of service, it must first secure from the Commission a "certificate of necessity and convenience." The Commission is supposed to grant the certificate only when satisfied that the service in question is needed and cannot be conveniently provided in any other way.

Attorneys for cable companies have strongly urged the FCC

that the phone companies' entry into cable TV, and their discrimination against independent system builders in favor of leasebacks, is exactly the kind of situation that Section 214 was designed to prevent. On reflection, the Commission agreed. In June 1968, in a case involving the ramified cable TV activities of General Telephone Company of California, the FCC ruled that any phone company wishing to build a cable system for leaseback purposes, would first have to get a Section 214 certificate of necessity and convenience.

Requests for such certificates began to pour in. Some of them were challenged by cable operators. For example, New York Telephone Company filed a request for a Section 214 certificate to build and lease back a cable system in Hyde Park. An independent operator, Better TV, Inc., that wanted to build and operate a cable system in Hyde Park, protested. In its filing with the Commission the firm charged that the phone company had delayed granting it a pole attachment contract, meanwhile negotiating, and then signing a contract with, another operator who was willing to take a lease. The case has been set down for hearing.

However, many of the phone companies' Section 214 certificate applications have not been challenged, since they cover communities where no independent has staked a claim. In February 1969 the Commission instructed its Common Carrier Bureau to give a rubber-stamp O.K. to any unchallenged Section 214 applications. The action, endorsed by a majority of the commissioners, evoked a stinging dissent from Commissioner Johnson. "This Commission," he said, "continues to persist in making decisions with immense future impact on our society without the slightest information or consideration as to where we are going or what we are trying to achieve."

This well-aimed shot seems to have hit its mark. By April the Commission had a backlog of seventeen Section 214 certificate applications from phone companies, all of them requesting permission to lease the completed system to a company that was partly or wholly owned by the phone company making the application. At that point the FCC announced a Notice of Inquiry and Proposed Rule Making on a whole series of questions raised by phone company involvement in cable.

Many parties filed statements with the Commission in connection with the proceeding, including the Department of Justice, the National Cable Television Association, General Electric Company, AT&T and several other phone companies, the National Businessmen's Council, and Americans for Democratic Action. Several cable companies also made filings, describing their pole-attachment woes.

By fall, although the proceeding had not been completed and the Commission had issued no statements or rules, AT&T quietly decided that the jig on this issue was up. It wrote a letter to the FCC, promising to place no impediments in the way of independent CATV system owners who wanted to rent space on poles or in conduits. In December, General Telephone and Electronics Corporation, largest of the independent phone companies, did the same.

The Commission issued its report and its new rules on February 4, 1970. They prohibit phone companies from operating, either directly or through a company in which they have an interest, a cable system in areas where they provide service. Leasebacks to genuinely independent operators are still permitted, but the phone company must make it clear to a potential leaseback customer that, if he would prefer to build his own system, he will be permitted to use phone company poles or conduits at reasonable fees.

If energetically enforced, these rules will slow the phone companies' drive toward hegemony over the Wired Nation, but the real issues have only begun to be explored. Phone companies still have significant ownership in cable TV, and leasing to genuine independents will continue. "They failed to achieve monopoly," says Lewis A. Rivlin, a Washington attorney who has represented cable operators in a number of suits against phone companies, "but they will settle for as much as they can get."

With the monopoly threat made less immediate, long-range questions must now be studied. What should be the role of the telephone company in the total communications setup of tomorrow? Should there be one great communications network, of which phone service is an integrated part? Should there be two systems, of which the phone company would be one? Should there be a number of systems? Should competitive firms offer the same or similar services?

In seeking answers, narrow approaches, old prejudices, and even many current arrangements, will have to be set aside. New concepts of the public and the national interest must be formulated. For example, Harold J. Barnett and Edward Greenberg, professors of economics at Washington University and consultants to the RAND Corporation, urge in the Winter 1968 issue of *Washington University Law Quarterly* a consolidation of telephone facilities and the cable network. "It would probably be sensible to merge the present telephone system into the WCTV [wired city television] system. In such an event, there could be available substantial economies in installation, maintenance, and operation of the WCTV system."

Other solutions are being discussed, and some unlikely groups find themselves working together. At the FCC rule-making inquiry, the National Businessmen's Council and Americans for Democratic Action filed briefs that were similar in many ways. The crux of the current problem, both groups felt, is that cable systems have not been designated as common carriers. If they were so designated, the two groups asked, what is the objection to letting phone companies bid for their construction, along with all other potential builders, and granting construction rights to whatever party offers the best system and the best rates?

This obviously moves the discussion into deep waters. We approach the fundamental issues involved in creating the Wired Nation. But first we should view the situation from the state and local levels.

6 *Grass Roots Action*

Some of the power struggles over cable at the Federal level have already been described. At the state and local levels other things are afoot. In a number of states there is substantial disillusionment with the ability of municipalities to franchise and regulate cable properly, and state legislatures are considering a partial or complete takeover of jurisdiction. In Illinois the State Public Service Commission did so by decree in September 1971, stating that cable is a public utility within the meaning of the laws that the Commission administers. Heated rhetoric surrounds this question of whether cable should be classified as a public utility, but at the bottom of it are issues of real importance.

Among other things, public utility status means that a state commission receives the authority to set standards of service and performance for cable companies. It also receives the power to regulate the rates that cable companies charge to subscribers. It is this latter power to which the cable industry is especially opposed. It advocates, instead, regulation of cable at the municipal and Federal levels, with the states playing no role at all. For cable companies this arrangement has an additional merit: it would create a regulatory framework in which one of the regulating parties is notoriously ill-informed and weak, and the other is too far away to exercise any kind of extensive ongoing supervision.

There seems little doubt, however, that, despite the wishes of the industry, the role of the states in cable regulation will increase, and the tendency to treat cable as some type of public utility will also increase. In August 1971, following the handing down of indictments alleging bribery in connection with cable franchising in Trenton, New Jersey, and allegations of corruption in other municipalities made during New Jersey state legislative hearings on cable, Senator Harrison Williams of New Jersey introduced Federal legislation that would establish cable TV as a public utility.

At the municipal level, the frequent failure of local governmental units to handle cable franchising properly for the benefit of their citizens, has already been described. Treatment of cable franchises as revenue bills, or as plums to be granted to favored friends or parties, has been frequent. Municipal officials who have been genuinely interested in learning about cable before framing any legislation have often found themselves in a minority in the city council or city administration.

The picture, however, is improving, and some of the improvement is being brought about by local civic action. Individuals, and local civic, educational, and religious groups, are increasingly playing a role in the franchising process. They can also play a role in bringing about creative civic, community, educational, and cultural uses of the cable after the franchise is granted. An interesting example of civic action recently occurred in Nashville, Tennessee, where readers of the *Nashville Banner* on April 2, 1971 learned some things that the city fathers had neither announced nor publicized. On March 16 a bill had been introduced into the city council, bearing the signatures of twenty-six of the council's forty members, to grant a cable franchise to wire up the city of

Nashville, to a firm called Mid-American Cable Corporation. The franchise was to be a forty-five-year exclusive grant to Mid-America, with no other company having the right to operate or compete with Mid-America during the entire term of the franchise. The city council planned to grant the franchise without competitive bidding and without drawing up any standards or requirements for the operation of the system. The bill was slated to move quickly to a second and third reading, and to pass before the upcoming city elections on August 5, at which time a substantial turnover in council membership was expected.

Most of the councilmen who sponsored and/or voted for the bill knew little about cable; many had gone along when asked to do so by those interested in the bill's passage. No council member seemed to feel that the council should take steps to become better informed, that it should seek expert assistance and advice, or that it should tell the community what was occurring and solicit its views.

Actually, there seemed little reason to think that the company itself knew a great deal more about cable than the city council, other than that cable has proved to be an excellent form of investment. Chairman of the board of Mid-America Cable Corporation is Chet Atkins, the famed country music performer. President of the firm is Owen Bradley, often called "Mr. Music Man," who is executive producer and vice president of Decca Records. The firm's other officers are prominent persons in Nashville's country music recording business.

The *Banner*'s revelations caused the city council some embarrassment, and it announced that it would be willing to receive competitive bids. Nine bidders expressed interest, six of whom filed formal applications for the franchise. Of these, five were profit-making companies and the sixth was the Nashville Metropolitan Education Association. After surveying the situation, NMEA decided that the best way to influence the course of events was to become an active contender for the operating rights. Along with its franchise bid the Association mounted an intensive campaign of civic action and public education.

NMEA educated itself on cable by seeking information, publications, and advice from national nonprofit organizations concerned with cable, and by consulting with nationally-known experts. It brought several experts to Nashville to hold press conferences and to testify before the city council. These experts described the many benefits that cable can bring to cities, and stated that the council was putting the cart before the horse. First, the experts said, the council should seek competent advice leading to the drafting of an ordinance which would set forth minimum standards and requirements for *any* company seeking the franchise. *Then* it could entertain competitive franchise bids, hold public hearings in which particular attention would be paid to community groups and their representatives, and select the bidder who offered the most to the city and the community.

Using the materials and information it had secured, NMEA prepared a list of sixteen "Possible School Uses of CATV" and twenty-three "Possible Municipal Uses of CATV" and distributed it to the city council, the press, and the public. Personnel from the Association went to the city records office, studied all the franchise bids, prepared a comparative chart showing what each bidder was offering in such categories as contract length, two-way capacity, expansion capability of the system, and channel capacity for civic, governmental, and educational use. The chart was supplied to the council and distributed to community groups and the public.

In the franchise contest itself the Association drew on expert counsel to draft a model ordinance under which it proposed to operate the system. Among its other provisions, the ordinance specified that, of the system's proposed twenty-four-channel capacity, five channels should be assigned for educational use, five for use by the municipal government, and five for use by the public, far more than the other bidders were offering or suggesting. It also proposed the shortest contract length—ten to fifteen years, compared with twenty to forty-five years proposed by the private bidders. MNEA's proposal would leave the city free at an earlier date to reconsider the whole franchise picture when the potential of cable is more fully developed.

The Association's activities soon attracted support. Among those who endorsed either MNEA's application for the franchise or its efforts to get the city council to set standards for franchising before entertaining applications, were the Metropolitan Board of Education, Vanderbilt University, the Television, Radio, and Film Commission of the Methodist Church, and leaders of the city's PTA's.

On June 21 two committees of the city council, the Judiciary and Legislative committee and the Budget and Finance Committee, with a combined membership of twenty councilmen, scheduled joint hearings to receive testimony from the franchise applicants. It quickly became clear which way the wind was blowing. In addition to the Chairman of the Judiciary and Legislative Committee who was chairing the hearings, only two committee members showed up at the June 21 hearings, and only six showed up at the continuation of the hearings the next day. The Chairman, stating that further hearings were pointless under the circumstances, turned the matter back to the full council, which immediately moved to reject the drafting of an ordinance creating standards for franchising, moved also to eliminate six of the seven applicants, and proceeded to approve a second reading of the bill giving the franchise to Mid-America Cable Corporation.

The Nashville Metropolitan Charter requires that franchise bills must be sent to the City Planning Commission for study and recommendations before final action on such bills can be taken by the council. On July 8 the Council sent all the franchise applications to the Planning Commission, requesting a response within a week. On July 15 the Planning Commission wrote to the council stating that it had conducted as intensive a study as time had permitted. The issues and problems, the Commission said, were complex; there were a host of new developments in the field; and the interests of the government and citizens of Nashville were profoundly involved. Cable systems will have an impact, now and in the future, on numerous aspects of urban life, including education, health, pollution, transportation, crime prevention, government and business administration, and citizen-government interaction. "The community," said the Commission, "does not seem to have the tools and information to act in its own best interests at this time." The Commission recommended that the council take no action on any of the franchise applications, and that instead the Commission and the council should begin intensive studies to develop guidelines and standards to be applied to any bidder seeking the franchise.

A week later, on July 20, the Council proceeded to the third reading of the bill and, by a vote of 27 to 11, granted the franchise to Mid-America Cable Corporation.

In December 1970 Mayor Henry Meier of Milwaukee became the first U.S. mayor to veto a cable franchise. Meier rejected a franchise grant that the city council had made to a firm whose officers were all local people but which was in reality a subsidiary of Time-Life Broadcasting Corporation. In his veto message Mayor Meier told the council that, in his opinion, it had not studied the issues sufficiently to ensure that the franchise gave the people of Milwaukee everything that they had a right to expect from the coming of the cable to the city. When the Mid-America franchise came to the desk of Nashville Mayor Beverly Briley on July 23, he had at least one precedent on which to reflect. He had other matters to ponder, too. Municipal elections, including the election for mayor, were scheduled for August 5. The franchise clearly represented an extraordinary union of political and economic power in the city.

After conducting intensive fact-finding with his staff, and being alone with whatever thoughts a mayor may have on such occasions, Briley bit the bullet on August 3 and vetoed the franchise. "I am sending this ordinance back, vetoed by me, in the public interest," he wrote to the council. When the matter is reconsidered, he added, it would be desirable that "criteria developed by the best technical assistants we can get . . . be utilized in the development of the ordinance, with the hope that local citizen participation can assume and be responsible for carrying out this type of criteria. I sincerely hope that your good judgment will agree with the best experts in the nation who have no particular interests with any of the CATV firms."

The next day the city council failed by one vote to override Briley's veto. The vote was 26 to 12, with 27 needed to provide the necessary two-thirds.

Events after that were swift. In the August 5 elections neither Mayor Briley nor his two leading opponents received a majority, but Briley won the subsequent runoff. During the campaign the Metropolitan Nashville Education Association published lists showing how each councilman who was running for reelection, had voted on the cable franchise. A number of them lost their seats, and the turnover in council membership was more than 50 percent. When the new council convened, the picture was transformed. By a vote of 30 to 9 the council deferred action on all the franchise applications, and returned the cable problem to the Planning Commission for the development of comprehensive standards and guidelines.

With its main objectives of public education and official action leading to the development of guidelines accomplished, MNEA decided to withdraw from the franchise competition. It plans to watch developments, to reenter the arena if events appear to warrant, and to participate in developing civic and educational uses of the cable when the franchise has been granted.

MNEA's campaign could serve as a model for groups seeking to influence the course of franchising in their cities and communities. First, it educated itself, seeking information from established national sources. Some of those sources are listed at the end of this chapter, and in appendix A. It sought out experts, brought them to town, secured help and information from them, scheduled press conferences for them, and arranged to have them appear before the city council. It studied the existing franchise bids and compared them with information, advice, and suggestions obtained from the sources and experts it had consulted. It urged the city council to seek its own expert advice and to set standards before attempting to consider franchise applications. It shared its information and ideas with other civic and community groups, and secured their interest and support. And, when there appeared to be no other effective avenue for securing city council consideration for a well-drafted franchise, the Association entered the bidding itself and submitted a model franchise drawn up with the assistance and cooperation of experts.

In Nashville this well-planned and well-executed effort was decisive in molding the course of events, in the face of an unusually close working relationship between the franchise seeker and the city council. In many other cities, civic and community groups can realistically hope for as much or more.

MNEA decided to drop out of the franchise race when it felt that its goals were being achieved, but civic and community groups in other cities may wish to bid for the franchise and stick it out to the end. The key problem, of course, is financing. The membership of community groups, and also of religious groups, often includes leaders of the local business and finance community, who should be consulted as a first step after the group has decided to become an active franchise seeker. The desirability of having at least some city franchises in the hands of nonprofit civic organizations, or in the hands of joint venture companies composed of a civic group and a profit-making partner, is a challenging concept for which financing will not be impossible to obtain.

In cities and communities where cable franchises have already been granted, civic and community groups can take the same first step in these communities as MNEA took in Nashville—seek information and materials from knowledgable sources, and educate themselves. Action can then proceed along several lines.

In many established systems, the cable operator has already opened up an empty channel for "cablecasting." Most operators are more than happy to confer with civic organizations about possible programming for this channel that the civic groups could initiate and create.

Most existing systems have *not* yet opened up or set aside channels especially for educational and civic use. A primary target for civic action in such communities, is to get such channels opened up and put into operation. Conferences can be held among cable operators, civic leaders, and city officials to secure agreement on the achievement of this goal even though no formal requirement for such channels exists in the franchise. Because shortsighted official action may have resulted in the building of a system whose expansion will involve additional cost, the problem of who will pay for what will have to be discussed and negotiated.

In larger cities and communities, if the franchise does not call for the creation of one or several neighborhood cablecasting studios, negotiations for this addition to the system's capabilities can be arranged and conducted by neighborhood groups.

* * *

Until now, city governments and civic groups have both been faced with a shortage of useful information to assist them in the franchising process and in making creative use of cable facilities. This situation is rapidly improving. In addition to the publications listed in the bibliography at the end of this book, other important information is becoming available.

——Foundation-supported studies are now being made in Washington, D.C. and Dayton, Ohio, preparatory to creating plans for the wiring of both cities. Information in these reports will be tailored to the individual requirements of the respective cities, but much of it will be of interest to other cities. Inquiries regarding the Washington study should be addressed to Mr. Forrest Chisman, The Markle Foundation, 50 Rockefeller Plaza, New York, N.Y. 10020. Inquiries regarding the Dayton study should be addressed to Mr. Stuart Sucherman, Ford Foundation, 320 E. 43rd St., New York, N.Y. 10017.

——The Ford Foundation has established a Cable Information Service, to provide counseling service on cable franchising to municipal governments. The Service is located at the Urban Institute, and inquiries should be addressed to the Urban Institute, 2100 M Street, N.W., Washington, D.C.

——The two franchises granted to cable companies in Manhattan, are the first city franchises that require the cable companies to provide channels that will be open to any and all members of the public. These public channels were formally opened on July 1, 1971, and use of them is now being organized. A civic group called Open Channel has been created to help civic groups and individuals make effective use of the channels. Information and suggestions on the use of this type of cable channel can be obtained from Mrs. Thea Sklover, Executive Director, Open Channel, 415 Madison Avenue, New York, N.Y. 10017.

Finally, it should be remembered that the potential of cable for civic, community, educational, and cultural use is only beginning to be explored. There is almost limitless opportunity for creative action at the community level, that will contribute to making cable communications a valuable addition to urban life, and a tool for dealing with the problems and ills facing our nation and our society.

Cable also poses important challenges for the nation as a whole, and it is to these that we will now turn.

7 *An Electronic Highway*

In the 1960s, the nation provided large federal subsidies for a new interstate highway system to facilitate and modernize the flow of automotive traffic in the United States. In the 1970s it should make a similar national commitment for an electronic highway system, to facilitate the exchange of information and ideas.

As in the building of any highway system, planning is necessary, and the nation must have some concept of what it wants. Inadequate or shortsighted approaches can result in large investments in systems that will not provide a full range of desirable or necessary service. This will mean either that the nation will do without such services, or that systems will soon have to be rebuilt at enormous additional cost.

There is still enough flexibility in cable TV to create national electronics highways in accordance with any pattern that the nation may select, but if planning is delayed, and if federal and state policies are not created to turn plans into reality, short-term commercial considerations will dictate the form of the network. A wired nation so created will almost certainly fail to incorporate services that would be strongly in the public and national interest—just as the present broadcasting system has failed from lack of national planning and policy. Unfortunately, no branch of the federal government has evolved such planning, nor is any in the process of doing so. The time when effective national decisions can be made and implemented is running out.

When the FCC asked in 1968 for comments on its new proposed rules for cable TV and on the future of the cable in U.S. communications, it received more than two hundred submissions. Most of them, however, discussed the short-range problems covered by the proposed rules.

A notable exception was a forty-page presentation entitled "The Future of Broadband Communications," filed in October 1969 by the Industrial Electronics Division of the Electronics Industry Association. "The Industrial Electronics Division, Electronic Industries Association (IED/EIA) views the services to be provided by broad-band communication networks in the late seventies and early eighties of landmark importance," says the report. "We look upon such systems as being of 'national resource' dimensions and the development of these resources as a national goal. . . . The mushrooming growth in available information and the demand for access to this information is bringing about a revolution in communications which will produce a profound change in the way society is structured and in the way we live."

"The Future of Broadband Communications" explains with a fair degree of clarity the major services and benefits that a national cable network can provide; describes the type of cable setup that would have to be built to provide such services; and speculates on the relationship of wired nation services to the society of the country. It therefore addresses itself to the central questions on which U.S. communications policy must almost immediately be based. To assist in this policy making, there should of course have been available by now a detailed study of these questions from a Presidential commission, the

FCC, the Congressional committees that have been dabbling in the short-range financial aspects of CATV, or from a source such as the RAND Corporation, which has been looking into cable TV with a generous Ford Foundation grant. But in the absence of any such study from any public or noncommercial source, IED/EIA's report is the best thing in its field.

IED/EIA, little known to the public, is highly regarded in Washington where it maintains national headquarters. Members of IED or its present association include IBM, Litton Industries, General Electric and other firms that would be the principal suppliers of the computers and other hardware needed to create the Wired Nation. During the 1960s, the association maintained gentlemanly silence on the industrial and commercial wrangles occasioned by the coming of the cable, a silence that may have been inspired in part by the fact that some of the equipment manufacturers in the membership are also in the broadcasting business. But it became obvious after a time that its interests did not coincide with those of the broadcasting industry. Broadcasters stood to lose from the diversity that the full development of the cable's potential would bring; IED/EIA's members stood to gain greatly from that development.

The political importance of "The Future of Broadband Communications" lies in the fact that IED/EIA is the first major industry group to break ranks with both the broadcasters and the cable TV industry, and to call for an immediate national commitment to wire the country. IED/EIA is also one of the few lobbies in Washington that may be a match for the broadcasters.

In the system proposed by IED/EIA, the nation's phone systems and its cable network would not be combined. The former would be a somewhat more elaborate version of what now exists, a system offering "videotelephone" service, with enough capacity to provide a limited amount of keyboard access to computers and to reproduce documents and other printed material in facsimile. These other services would be optional, and would probably be used principally by business, as is the case today.

The second system would be the broad-band cable network (BCN). It would provide a highway for national transmission and exchange of visual material, facsimile printed material and data. Each installation would include a narrower "return bandwidth," which would allow the subscriber to "talk" to central switching mechanisms, data storage banks and computers, to tell them what he wants to see or to have transmitted in facsimile. "We are talking," says the report, "of moving, at low cost, large amounts of information from central sources to the individual plants and homes where it is needed."

The report discusses a number of functions that such a system can perform:

Home Library Services. The contents of libraries can be put on microfilm or video tape. Using his "narrowband return channel," the home or business user can request any book or publication. He receives it as page-by-page visual images on his screen, signaling for each page in turn. Or an actual copy of the book can be transmitted to him since, in addition to a visual screen, his equipment will include a facsimile recorder.

Facsimile Data Service. This service would provide business firms or individuals with what the report calls "selected visual material by category." It could consist of summary information on given subjects, or continuing summaries of new information or events relating to given subjects, or both. "The public acceptance has not been demonstrated," says the report, "since an equivalent service does not now exist to provide a measure of need. The reader will note the parallel to the case of the telephone demonstrated by Bell at the Philadelphia

Centennial Exposition in 1876; no one then saw any need for such a device."

Delivery of Mail. The volume of paper work generated by business and private sources is expected to more than double in the coming decade. "This mass of paper," says the report, "will be sufficient to block communications channels as they now exist."

One institution that will probably be blocked for good unless something fundamental is done is the post office. It is estimated that by 1980, 108 billion pieces of mail (*excluding* checks!) will be handled annually.

An analysis of the mails made in 1968 by Arthur D. Little, Inc., for the President's Commission on Postal Organization, produced the following breakdown:

Transactions (checks, money orders, bills or statements, purchase orders, or questions about orders)	40%
Advertising	26%
Correspondence (personal, 13%, business, 5%, government, 4%)	22%
Magazines and newspapers	11%
Merchandise	1%

All this mail except the merchandise—that is, 99 percent of all mail—can be transmitted by cable. The largest single category, the 40 percent that represents transactions, is particularly well suited to cable transmission. Ordering, billing and payment of bills and accounts can all be handled automatically by computers that keep each person's account, and receive appropriate data by cable from buyers and sellers. "The financial community estimates that before 1980, we should be conducting almost all our fiscal operations, business and private, without the interchange of money or paper."

Correspondence can easily be sent by cable, once the appropriate system has been created, and letters will arrive at their destinations, anywhere in the United States, within minutes after they are posted. Third-class mail and advertising material can be sent simultaneously to coded groups of recipients, in accordance with the mailer's wishes.

Crime Detection and Prevention. The ability of law-enforcement agencies to deal with the growing incidence of crime may decrease steadily during the coming decade unless new detection and prevention techniques are developed. Foremost among these are the services that can be offered by a cable system. Within a community, streets and stores can be kept under surveillance from a central source. On a national level, computer-stored information can be available on a uniform basis to all law-enforcement agencies.

Travel. By 1980 physical travel, like the movement of paper, will have reached the saturation point. Cable networks will substantially reduce the need for business travel. Information and documents can be instantly obtained from distant sources. Conferences can be held by videophone, with each person able to see the others, and with all parties having access to documents.

The housewife can avoid trips to local stores and shopping malls. She can request displays of merchandise and prices from various stores on her home screen, and place her orders on the return band of the cable. The store will transmit the bill to the central accounting computer, which will automatically transfer the funds from the family's balance to the store's account.

Not all sources and authorities will agree with all of IED/EIA's concepts of the type of system that should be developed, or how it should be developed, but the organization is nevertheless the first to come up with a comprehensive working plan for the Wired Nation. The report makes clear that many details will have to be refined through experience, but it provides the description of a specific model, all of whose features are within easy reach of existing technology, and urges the necessity of reaching decisions and making a national commitment.

Important as the IED/EIA report may be for its pioneering description of a workable wired nation, it may prove even more important for two densely worded sentences that appear on page 32: "The public interest can best be served currently, in the immediate future, and the long run, by preservation of free, open and effective competition for the communication services that are and will become economically feasible where broadband cable systems are installed. To insure this development, there should be a clear distinction between the communication service (CATV, video telephone, mail and data communication, etc.) being offered the public, and the transmission medium (broad-band cable systems) which carries the communications service." The words "common carrier" do not appear, but the meaning of this passage is unequivocal.

As with the National Businessmen's Council, the logic of its own commercial interests has caused IED/EIA to take a position exactly opposite to that embraced by the broadcast industry and the present cable industry. It lines up with such groups as Americans for Democratic Action and the American Civil Liberties Union in advocating that cable system be stripped of control over content. This is tantamount to saying that they should and must be common carriers.

If the CATV systems are designated common carriers, the owners will have no stake in any of the programs and services; rather, their interest will be to achieve as great a proliferation of programs and services as possible. For the National Businessmen's Council, this means competition on rates for the services they will want to use and, probably, a wider selection of such services. For IED/EIA, it means sale of more hardware. "With all due respect for whatever clout ADA and ACLU may have on the Hill," says one knowledgeable observer, "EIA's view is likely to be the thing that will turn the trick."

Actually, cable TV is, at present, not only incorrectly set up to provide full benefits to the public; it is set up in a way to abridge basic freedoms of speech, press and assembly. If action is not taken by Congress, these characteristics will become basic features of the Wired Nation.

The defects of the present operation can best be understood by breaking the industry down into its three basic categories: the companies that manufacture the equipment and supplies used to build cable systems; the cable systems themselves; and the TV programs and other services that are sent over the cable.

A moment's reflection makes it clear that none of these three entities should be allowed to engage, or have any direct economic interest in, the activities of the other two.

No economic link should exist between cable companies and equipment suppliers. Cable companies should be free to buy equipment on the bases of quality, service and price. They should not be the captive customers of any company, or obliged to consider the welfare of a parent or subsidiary company.

Cable companies should exercise no judgment on the TV programming or any other service that goes over their cables. They should not be tempted to evaluate the content of such programming, or to limit offerings that might compete with their own programs.

If the cable operator is limited to leasing channel space at standard rates to all comers, then, of course, the cable becomes a common carrier; its nature and functions are transformed. *Congress should legislate the separation of the three entities in the cable industry, and should designate individual cable systems, and combined cable networks as common carriers, with access guaranteed to all comers.*

In addition to being made common carriers, cable systems should be assigned the status of public utilities. That would place them under additional controls and obligations. First, they would have to maintain certain standards of performance and a certain quality of service, as specified by the FCC and state Public Service Commissions. Second, they would have to conform to certain uniform technical standards, which would make possible interconnection of systems. Third, they could not do what the broadcast industry has grown rich by doing, and what the cable systems are doing now—provide service only in communities where high profits beckon. As public utilities, cable systems would be required to provide equal service to all communities. In exchange, the cable system would be permitted to average its total profits and losses for rate-making purposes. Such regulation, the City Club of New York told the New York State legislature in 1969, "can abolish second-class citizenship in communications without tangible additional costs to either the public, broadcasters, or to the cable carriers." And last but far from least, if cable systems were classified as public utilities, their rates could be regulated to insure fair tariffs and a fair return.

Cable systems meet every historic test for classification as common carrier and regulation as public utilities:

They are natural and unavoidable monopolies. Few communities need, or will support, two cable systems.

They provide essential service. Access to a full offering of diversified television communications and supplemental electronics services is not a luxury. It is probably, as IED/EIA strongly maintains in its FCC filing, an essential condition for the survival of our society.

A full offering of such facilities must be uniformly and universally available for public needs and public use.

The desirability of making at least part of the channel capacity of cable systems available on a common-carrier basis is now being discussed in many quarters. For example, in its 1968 and 1971 proposed rules, the FCC gave cable systems permission to devote one or more channels to common-carrier service. That recalls Sidney W. Dean's comment, quoted earlier, that if the FCC had made common-carrier service mandatory and program origination optional, "we would be in the right ball park at last."

Dean's view was echoed in more formal terms by the American Civil Liberties Union. "The only basis upon which the cable system can develop into the two-way switched public utility system for which it has such great potential," the ACLU said in its filing with the FCC, "is through its operation as a common carrier. . . . Yet it is here that the Commission folds its hands and complacently leaves to operators the option as to whether and to what extent it will make leased channels and time periods available on a common-carrier basis."

In addition to freeing program content, and decisions as to the types of services that the cable will carry, from the control of the cable operator, common-carrier status would end a tremendous denial of basic freedoms that has been inherent in American TV. No provision exists, or ever has existed, for public access to broadcasting facilities, exclusive control resting with a small group.

It is not easy to understand why this way of disposing of the publicly owned electromagnetic spectrum has never been challenged in the courts as a violation of the First Amendment. In its brief, ADA wasted no words in setting forth its view. "Without affirmative safeguards by the Congress, electronic technology under laissez-faire economic dominance has denied the rights of the people to unabridged speech, press and assembly in the dominant media of our generation."

Cable systems would probably not oppose a requirement that they make one or more of their channels available to all comers, but the industry is fighting any move to designate cable systems in their entirety as common carriers. And it is battling as vigorously to escape public utility classification and regulation—especially of rates—either by FCC or by the states.

Rhode Island, Connecticut, Nevada and Vermont have passed laws bringing cable systems under the regulation of the state Public Service Commission. Many Connecticut communities need CATV, and before passage of the new law, numerous operators had applied for franchises. Since then, no cable system has been built in the state. The operators have gone elsewhere, where their profits will not be scrutinized or regulated. By doing so, they prove the importance of adopting public utility regulation in all the states.

In Nevada, two CATV firms went to court to escape regulation, claiming that, because the FCC had asserted authority over cable television, the federal government had pre-empted the field, and states were powerless to initiate public utility surveillance. A federal court dismissed the suit, stating that federal and state regulatory authority did not overlap; in February the Supreme Court upheld the decision.

Not only must charges for new and special services be appraised by public authorities but the fairness of existing charges also needs review. Most cities and communities grant franchises without having a background of information or experience to make certain that proposed monthly charges are fair. In addition, as noted, many municipalities have not cared about fairness—as long as the public purse received a big slice of the profits. Now monthly charges for cable rental are creeping up. *CATV,* the industry magazine, contains small news items such as:

> *Pennsylvania:* Milton CATV, Inc. has announced a monthly rate hike of $1 for their subscribers in Danville, Lewisburg, Milton and Watsontown. Milton CATV offers 7 video channels.
>
> *Wisconsin:* Abel Cable Co. has announced a monthly rate hike of $.50 for subscribers in Eau Claire. The firm offers subscribers 10 video channels.

This trend contrasts with calculations made by Barnett and Greenberg (*Washington University Law Quarterly*). Reviewing estimated capital cost statistics in a number of towns and cities of various sizes, the authors conclude that "the cost of wiring a community varies between $30 and $80 per home. At 15 percent per annum for capital charges and maintenance, this would calculate to $1 per month or less. Allowing for contingencies and uncertainty, a figure of $1.50 is possible." The authors caution that the figures are "very rough" but they are sufficient reason for a thoroughgoing review by competent authorities of the whole structure of cable television charges. The matter will become increasingly important as cable largely or wholly supplants over-the-air broadcasting in major urban areas, when virtually everyone will be required to pay.

An important corollary is that the basic public utility functions of setting uniform technical specifications, and determining fair rates, must be retained at the federal level. This is the exact opposite of what the industry wants, and the exact opposite of current FCC policy, which holds that these and similar problems should be handled by local authorities.

How would the television part of the Wired Nation operate? Apart from those channels devoted to the retransmission of established commercial and educational stations, all channels would be available at regulated, uniform rates to business, government, civic, fraternal, labor, religious and other groups, and to individual members of the public, in time segments of, say, fifteen minutes and up. In addition to the actual cable, the common-carrier system could provide studios, cameras, tape recorders and players, studio personnel and related facilities, which the person leasing time might use as he wished.

The cost of renting time on the cable might turn out to be remarkably low. Barnett and Greenberg estimate that it would cost $6 million to wire completely a city of 100,000 homes (approximately 500,000 people). "[If] the common carrier were to charge . . . a rental of 5 percent of the cost of the wired city each year, then the cost to senders of TV signals . . . would be $300,000 per year. This is less than $50 per day per channel. For a broadcast day of ten hours, this is less than $5 per broadcast hour." The difference between costs in this range, and the price that, for example, a candidate for office must now pay for air time, needs no comment.

How much will it cost to create the Wired Nation? For purposes of comparison, the electric power system operates with some $7 billion in gross plant investment, and the telephone system with some $51 billion in gross plant investment. Several guesses have been made for the national cable network, these assuming that the more available half of homes in America could be wired up for a given amount, but that costs would rise sharply as one tried to push the cable into every isolated corner of the country. A vice president of Bell Laboratories stated recently that it would cost about $12 billion to provide twenty-channel cable service to the more accessible 50 percent of American homes. Beyond that midpoint, he did not go.

A study prepared for the President's Task Force on Telecommunications by Complan Associates estimated that there will be 100 million homes in the United States by 1980, and that about half of them could be wired up for about $5 billion. However, to reach them all would cost $123 billion. This, however, neglects the proved feasibility of servicing inaccessible areas and homes by over-the-air microwave relays and "earth satellites," instead of laying a cable to every last remote door.

ADA has also made an estimate: "Capital requirements for the basic carrier system are extended in the range of $15 billion to $40 billion, dependent upon the degree of integration with present telephone and telecommunications systems and the provision of subscriber terminal equipment." ADA believes that most of this sum can be raised, over a period of five years, from private investment sources, assisted by federal loans or guarantees. For comparison, it notes that the Federal Interstate Highway System has received more than $50 billion in outright federal grants in the past ten years.

It is hard to assign a dollar value to many or most of the educational, cultural, recreational, social and political benefits that the nation would receive from a national communications highway. It is easier to assert the negative—that the nation probably cannot afford not to build it. But some interesting estimates have been worked out in areas where dollar values can be calculated.

IED/EIA believes that its proposed videophone-BCN network would result in immense annual savings. It summarizes as follows:

Domestic air travel	— over	$6 billion
Highways	— over	$6 billion
Police protection	— over	$3 billion
Fire Protection	— over	$1 billion
Post Office	— nearly	$6 billion
Recreation	— over	$28 billion
Total	— over	$50 billion

The City Club of New York stated in testimony before the state legislature that communications economists foresee public and business purchases of broad-band communications services, once the network is completed, of between $40 billion and $80 billion a year. If 10 percent of this total were spent in New York State, the City Club observed, and if it were subject to the usual franchise tax of 5 percent, the annual revenue to government units in the state would run between $200 million and $400 million.

All these figures are rough, and they do not agree; but this much is evident—the network for a wired nation is well within the capacity of the economy. It can be done, and in one fashion or another it undoubtedly will be done.

Therefore, a number of important questions need to be answered, and soon. An agreed upon overall plan is needed, and reasonably accurate cost figures must then be developed for it. The answer to the problem of remote homes may lie in the creation of a Rural Communications Administration, similar to the Rural Electrification Administration. Another serious question is how the nation's poor will pay for the new service, if and when it supplants, wholly or in part, the present system of free over-the-air transmission. The difference between current charges of $5 and up per month and the $1 to $1.50 figure of Professors Greenberg and Barnett, is fundamental for public policy in this area. A subsidy for the poor, if it proves necessary, cannot be shirked.

There are yet other issues. It cannot be assumed that all the social effects of the cable will be good. For example, the exodus of the middle and upper-middle class from the cities is expected to continue during the 1970s, and the stratification of society along geographic-economic lines will thereby be increased. At the same time, the cable will make it less and less necessary for the more affluent population of the suburbs to enter the city, either for work or recreation. Lack of concern and alienation could easily deepen, with effects that could cancel the benefits of community expression that the cable will bring to inner-city neighborhoods. At the very least, such dangerous possibilities must be foreseen, and the educational potential of the cable itself must be strongly marshalled to meet them. The bland treatment of this issue in the IED/EIA report is chilling.

There are other serious problems. For example, in several cities and communities, if one stands at a busy intersection and happens to glance up at a nearby utility pole, one will see a television camera moving slowly from left to right, scanning all persons and events at the intersection. This camera is connected by cable to police headquarters, where a single policeman, sitting in front of TV monitors, can perform the work of a number of policemen on the beat in overseeing the midcity area.

Efficient? Yes. Desirable? That's harder to answer. When cable systems blanket major cities, cameras can be put everywhere. Everyone's comings and goings can be monitored every minute and hour of the day. Will this new form of police power be properly used? Should it be limited, and, if so, what should the limits be?

Cable poses other invasion-of-privacy issues. For example, even with present technology, a cable operator can send a pulse through his system and tell, at any hour of the day or night, which of his subscribers is tuned to what channels. Efficient for market research? Yes. A desirable kind of information on one's private life to be compiled by central sources? Many people think not.

The compiling of financial, credit, and other personal information about individuals in computer banks, which would be available instantly by cable to all persons possessing or knowing the correct code, raises unprecedented complexities of civil liberties and privacy. Privacy problems are also involved in the transmission of mail by cable. And in the creation of Facsimile Data Services such as those envisioned by IED/EIA in its report, the question is, who will decide what data is to be included in these services, and what is to be left out? Unless the issues in these future uses of the cable are understood and faced, 1984 could easily come well in advance of George Orwell's prediction.

Under any circumstances, however, the cable will be built, and the aim must be, through positive policy and intelligent action, to take every advantage of its tremendous potentials for social good. The way forward now is through the establishment of a Presidential commission on the Wired Nation. It would have as its mandate the development of a plan for creating a national broad-band communications system in the United States during the 1970s, and should be asked to submit its report within a year. That commission should conduct extensive consultations with economists, the business community, communications technicians and social scientists. Its report and recommendations might well serve as the basis for a special Presidential message to Congress, setting forth the plan, describing the type and amount of recommended federal participation, and seeking full Congressional and public support.

There is no reason to think that such support would be denied. But further delay at this juncture will forfeit great opportunities for shaping the swiftly approaching communications revolution for the public benefit. Leadership is required, and the time is now.

APPENDIX A

Bibliography

Mr. Jon Shafer, a graduate student at the University of Minnesota, has interested himself in the cable problem, and has become heavily involved in civic activity relating to cable in the State of Minnesota. In the summer of 1971 he compiled and circulated an annotated bibliography on cable which the author of this book regards as the best and most useful that has yet appeared. With Mr. Shafer's kind permission, his excellent bibliography is reproduced below.

I. Reports, Papers, and Booklets. (Free unless otherwise noted.)

Better Broadcasting Council. *A Model Ordinance for Cable Television for the City of Chicago.* Chicago: Better Broadcasting Council, 1970. 15pp. $2.00.

A good attempt to write the ideal ordinance, from a public interest perspective.

Write: Better Broadcasting Council, 53 W. Jackson Blvd., Room 630, Chicago, Illinois 60604; or call: (312) 922–1372.

Center for Analysis of Public Issues. *Crossed Wires: Cable Television in New Jersey.* Princeton: The Center for Analysis of Public Issues, 1971. 94pp. $5.00.

A superior work, essential to those seriously interested in CATV. A superb evaluation of the extensive experience of New Jersey municipalities with CATV. Included are listings of the important issues in franchise negotiations, as well as a sample CATV franchise. The importance of avoiding hasty decisions is stressed.

Write: The Center for Analysis of Public Issues, 92A Nassau Street, Princeton, New Jersey 08540; or call (609) 924–9750.

———. *Public Issues: New York CATV Supplement.* 1971.

"In this series, we plan to inform potential users of developments in the New York City public access CATV channels, discuss problems in planning and telecasting public access productions, and highlight interesting and innovative uses of the channels." (Supplement No. 1 [July, 1971], p. 1.)

Write: New York CATV Project, Center for Analysis of Public Issues, 18th Floor, 415 Madison Avenue, New York, N.Y. 10017; or call: (212) 758–6017.

"Community Cable TV and You," National Film Board of Canada, *Challenge for Change Newsletter,* February, 1971, 12pp.

This cartoon-assisted introduction to CATV is a joy to read and guaranteed to communicate the basics of CATV to anyone. Subscribe to the *Newsletter;* it's free, fun, and relevant.

Write: Challenge for Change Newsletter, N.F.B., P.O. Box 6100, Montreal 101, Quebec, Canada.

Electronics Industry Association. "The Future of Broadband Communications: The IED/EIA Response to Federal Communications Docket 18397, Part V." October 29, 1969. A position paper prepared by the Industrial Electronics Division of EIA. Washington: Electronics Industry Assoc., 1969. 41pp. (Mimeographed.)

Broad-range long-term thinking from the electronics industry, covering many of the myriad uses of coaxial cable, e.g., electronic mail, banking, shopping, automated libraries, and all the rest.

Write: Public Relations Department, National Cable Television Assoc., 918 16th St. N.W., Washington, D.C. 20006; or call: (212) 466–8111.

Federal Communications Commission. *Community Antenna Television (CATV)*. Information Bulletin 22–D (November, 1970). Washington: F.C.C., 1970. 16pp.

A handy short history of CATV, and federal regulation of it.

Write: Federal Communications Commission, Office of Information or Cable Television Bureau, 1919 M Street N.W., Washington, D.C. 20554; or call: (202) 632–7260 or 6480.

———. Untitled. (A letter of intent to Congressional communications subcommittees outlining the forthcoming Third Report and Order.) FCC 71–787 63303 (August 5, 1971). Washington: F.C.C., 1971. 55pp and 18 attachments.

This letter is essential reading. It is the outline of the FCC's intentions for CATV, providing a regulatory foundation for years to come. A landmark document signalling the FCC's acceptance of CATV which, if put into practice as scheduled in March, 1972, will unleash cable into the nation's 100 largest markets with full force. Very strong on public interest matters of public access and 2-way requirement. It faces delay from friends of broadcasting in Congress.

Write: see above.

Ford Foundation. Comments of the Ford Foundation in Response to the Commission's Notice of Proposed Rule Making of July 1, 1970, in the Above Entitled Matter (FCC Docket No. 18892, Amendment of Part 74, Subpart K). New York: Ford Foundation, 1970. 32pp.

A well-reasoned argument for public ownership of cable systems. Also available are free copies of the 5 Rand Reports on CATV, which Ford funded. The most relevant is "Cable Television: Opportunities and Problems in Local Program Origination," by N. E. Feldman (see the Rand entry for others).

Write: The Ford Foundation, Office of Public Broadcasting, 320 E. 43rd Street, New York, N.Y. 10017; or call (212) 573–5000.

League of Minnesota Municipalities. *CATV*. A report prepared by the Metropolitan Section. Minneapolis: League of Minnesota Municipalities, 1971. 43pp. $3.00.

A short introduction to cable for municipal officials, including

25 pages of Bloomington, Minn.'s ordinance and pre-ordinance legislation. Not as incisive as SEMCOG's similarly-intended report. (See SEMCOG entry.)

Write: League of Minnesota Municipalities, 3300 University Ave. S.E., Minneapolis, Minn. 55414; or call (612) 373–9992.

National Academy of Engineering. *Communication Technology for Urban Improvement.* A report of the Committee on Telecommunications. Washington, D.C.: National Academy of Engineering, 1971. $3.00.

Ideas on how cable can help solve many urban problems.

Write: National Academy of Engineering, Committee on Telecommunications, 2101 Constitution Ave. N.W., Washington, D.C. 20418; or call (202) 393–8100.

National Cable Television Association.

Many items are available on different aspects of cable's past, present, and future. Most are light reading, but the historical summaries are useful.

Write: Donald Witheridge, Director of Public Relations, NCTA, 918 16th St. N.W., Washington, D.C. 20006; or call: (212) 466–8111.

National Education Association. *Schools and cable television.* A booklet prepared by the Division of Educational Technology. Washington: National Education Assoc., 1971. vi and 66pp. $2.25. 10% off for 2–9; 20% off for more.

Superb. ". . . the reason for this publication—to alert the teaching profession to the opportunities and options available in the development of cable television and to offer guidelines for the sound development of this new communications system in the public interest in the years ahead." (P. vi.)

Write: NEA Publications-Sales Section, 1201 16th St. N.W., Washington, D.C. 20036; or call: (202) 833–4231.

New York State Education Department. "Cable Television Information Packet." Compiled by Dr. Bernard Cooper.

An assortment of testimony, bills, and sample franchises focusing on the educational issues in CATV.

Write: Bureau of Mass Communications, Division of Educational Communication, Room 264, State Education Dept., Albany, New York 12224; or call: (518) 474–2121.

New York State Public Service Commission. *Regulation of Cable Television by the State of New York.* Report to the Commission by Commissioner William K. Jones. Albany. N.Y. State Public Service Commission, 1970. 220pp. $5.00. Mimeographed.

History and advice.

Write: Francis Rivett, Public Service Commission, 44 Holland Ave., Albany, N.Y. 12208; or call (518) 474–7080.

Philadelphia City Planning Commission. *Telecommunications Needs for Municipal Functions in Philadelphia.* Part I: *The Libraries.* Part II: *The Department of Public Health.* A staff paper by the Division of General Research. Philadelphia: Philadelphia City Planning Commission, 1970. Mimeographed. II/13pp.

This paper's purpose ". . . was to obtain some guidelines for the possible effect of telecommunications on the location and distribution of various public facilities in Philadelphia, among them branch libraries, neighborhood health centers, schools, and so on." More parts to come.

Write: Ralph B. Hirsch, Division of General Research, Philadelphia City Planning Commission, City Hall Annex, 13th Floor, Philadelphia, Pa. 19107; or call: (215) MU6–4620.

Radical Software. Issues 2–4 available. New York: Raindance Corp., 1970–71. #2, $1,25; #3, $1.50; #4, $3.00; (yet to come) #5, $3.00; and #6, *Guerrilla Television,* $3.95.

Informational journal of the video underground, including valuable information on cable TV, ½" video equipment, and details of experiments in community organizing, education, art, *et al.* It also lists everyone working in video who bothered to write them.

Write them: Raindance, 8 E. 12th St., New York, N.Y. 10013.

Rand Corporation. Reports, Memos, and Papers.

Good, but not great. The many and varied Rand materials cover a wide range of communications topics. They include Reports on CATV and: satellites, ombudsmen, education, urban development, local origination, broadcasting, monopoly problems, 1st amendment rights, and the future.

Read at the library, obtain 5 free from Ford Foundation, or write: Communications Dept., The Rand Corporation, 1700 Main St., Santa Monica, Calif. 90406.

SCOPE (Suffolk County Organization for the Promotion of Education). "CATV Briefing Packet."

Contains article reprints, papers, and sample cable TV franchises, all oriented around educational uses of CATV.

Write: SCOPE, Main St., Setauket, N.Y. 11733; or call: (516) 751–8500.

SEMCOG (Southeast Michigan Council of Governments). *A Preliminary Report Concerning the Problems and the Potential of Community Antenna Television, CATV, in Southeast Michigan.* A report by Patrick Kenney. Detroit: SEMCOG, 1971. 28pp.

An excellent report on the important issues to be considered, from a metropolitan and municipal perspective, in looking at cable communications and its tremendous impact. It suggests fully involving school systems and the establishment of a citizen's cable commission, with full regulatory powers.

Write: SEMCOG, 8th flr. Book Bldg., Detroit 48226; call: (313) 961–4266.

Singer, Arthur L., Jr. *Issues for Study in Cable Communications.* An Occasional Paper from the Alfred P. Sloan Foundation. New York: Alfred P. Sloan Foundation, 1970. 19pp.

Singer asks many of the right questions about the proper direction for cable communications. This is the only report currently available for general circulation, of the 50 or so done by, and for, the blue-ribbon Sloan Commission on Cable Communications. The remainder will be released approximately November 1, 1971, along with their Final Report, which should be excellent.

Write: Sloan Commission on Cable Communications, 105 Madison Ave., New York, N.Y. 10016; or call (212) 684–1800.

Strasser, Judith. *Why Wire the Nation?* Stanford, Calif.: Institute for Communication Research, 1970. 26pp.

A political analysis of the cable scene, arguing for resistance to division-producing technology such as CATV, until the benefits can be equally shared. Other material is also available.

Write: Edwin Parker, Institute for Communication Research, Stanford Univ., Stanford, Calif. 94305; or call (415) 321–2300 ext. 2755.

II. Articles and Periodical Indexes.

Alpert, Arthur. "Crossed Wires: Cable TV and the Public Interest," *Washington Monthly,* July, 1969, pp. 35–46.

An aged but informed history of cable, especially good on political analysis of developments in federal regulation and on the national power struggle over cable.

Asimov, Isaac. "The Fourth Revolution," *Saturday Review,* Vol. 53, No. 43 (October 24, 1970), pp. 17–20.

A well-reasoned overview of communications revolutions and their effects on civilizations, that contains much food for thought. We're in the fourth revolution now, i.e., developing instant 'any' communication through cable and satellites.

Barnett, Stephen R. "Cable Television and Media Concentration. Part I: Control of Cable Systems by Local Broadcasters," *Stanford Law Review,* Vol. 22, January 1970, pp. 221–329.

"The present Article, after providing rather extensive background, will focus on the issues raised by common control between cable television systems and television or radio broadcast stations serving the same market." (P. 222.) Easy to read, important, and fantastically well-documented.

Johnson, Nicholas. "CATV: Promise and Peril," *Saturday Review,* Vol. 50, November 11, 1967, pp. 87–98.

The Ralph Nader of the FCC has many worthwhile thoughts on cable. This article makes up the only cable chapter of his book, *How to Talk Back to your Television Set.* New York: Bantam, 1970, 95¢.

———. "What Do We Do About Television," *Saturday Review,* Vol. 53, No. 28 (July 11, 1970), pp. 14–16 and 34–35.

This is mainly about broadcast television public interest tactics, but there are 3 incisive paragraphs on cable on the last page.

Oppenheim, Jerrold N. "Cable TV comes to Clout City," *Chicago Journalism Review,* Vol. 4, No. 8 (August, 1971), pp. 3–14.

A picture of the politics surrounding Chicago's present cable television activity. A line up of their cable contenders in this pre-ordinance period.

Pool, Ithiel de Sola. "Social Trends," *Science and Technology,* No. 76 (April, 1968), pp. 87–101. A Special Issue on "The Communications Revolution."

"The new technology will tend to substitute for the mass media an interactive medium that can adjust to each user's desires and thus introduce into society a powerful force toward fragmentation and variety." (P. 87.) Comprehensive thoughts on the interaction of communications technology and society.

Tewlow, Jules S. "CATV in Perspective," American Newspaper Publishers Association Research Institute, *R. I. Bulletin,* No. 1029 (September 25, 1970), 36pp.

An excellent summation of FCC actions and considerations on CATV, plus an understandable review of technological advances like 2-way systems, video cassettes, laser, and satellites.

Write: ANPA Research Center, Box 598, Easton, Pa. 18042; or call (215) 253–6155.

Thorp, Bruce E. "Communications Report/FCC moves toward decision on rules vital to CATV industry," *National Journal,* Vol. 3, No. 1 (January 2, 1971), pp. 1–11.

A thorough, balanced review of recent FCC hearings leading up to their letter of intent on the Third Report and Order (see FCC entry), including the various Commissioner's viewpoints.

Vento, Charles J. "Making the Most of CATV," *Educational Television,* Vol. 3, No. 7 (July, 1971), pp. 9–10.

Good suggestions on how to make future CATV systems serve community needs.

White, Stephen. "Toward a modest experiment in cable television," *The Public Interest,* Summer, 1968, pp. 52–66.

Explains in easily understandable terms, the magnitude of the change CATV could bring about. Focuses on advantages of guaranteed hook-up and adequate access for an informationally-isolated community like Bedford-Stuyvesant in Brooklyn.

Indexes. For further reading.

Education Index. Few entries, but useful.

Look under: "Television Antennas—Multiple outlet systems."

Index to Legal Periodicals. A significant number of very significant articles, easy-to-read, and well-documented.

Look under: "Television—Cable Television."

New York Times Index. Good mainly for cable activities in New York and Washington, D.C.

Look under: "Television and Radio—Community Antenna Television (CATV)."

Public Affairs Information Service Bulletins. Perhaps the most fruitful.

Look under: "Television relay systesm."

Reader's Guide to Periodic Literature. Mediocre, but widely available.

Look under: "CATV System."

III. CATV Trade Journals and Reference Books.

For an inside view of the CATV industry. Unfortunately, few of these are available in the local libraries. Talk to your librarian.

Broadcasting Publications, Inc., 1735 DeSalles St. N.W., Washington, D.C. 20036:

Broadcasting. A newsweekly for both broadcasting and cable, especially good for FCC coverage. $14/yr.

CATV Sourcebook. Annual reference book. $8.50.

Broadcasting Yearbook. $11.50.

Communications Publishing Corp., 1900 W. Yale, Englewood, Colorado 80110:

CATV Magazine. "Authoritative Newsweekly of Cable Television." $33/yr.

TV Communications. "Professional Journal of Cable Television." $10/yr.

Originating Programs on Your Cable System: A Collection of Outstanding Cablecasting Articles from *TV Communications.* 1969.

CATV Directory of Equipment and Services. $6.95. Free with *CATV* subscription.

CATV Systems Directory. $8.95. Free with *CATV* subscription.

C. S. Tepfer Publishing Co., Inc., 607 Main Street, Ridgefield, Conn.:

Cablecasting. Monthly. $18/yr.

Educational Television. Monthly. $18/yr.

T.V. Digest, Inc., 2025 Eye St. N.W., Washington, D.C. 20006:

CATV & Station Coverage Atlas. Annual reference book. $17.50.

Television Digest with Consumer Electronics. Weekly. $95/yr.

T.V. Factbook # 40: 1970–71 Edition. Includes CATV statistics. $38.

Addenda

Since Jon Shafer compiled his bibliography, a number of new items have appeared or are scheduled to appear. Some of these are listed below.

Jurgen, Ronald K. "Two-way Applications for Cable Television Systems in the '70s." *Spectrum* (Magazine of the Institute of Electrical and Electronics Engineers), November 1971.

Mitre Corporation. *Urban Cable Systems.* An Interim Report Prepared Under a Grant from the John and Mary Markle Foundation. M71–6. McLean, Va.: Mitre Corporation, November 1971. 305pp.

A final version of this bulky interim report will have been published by the time that this book is published. Although it is a report on the feasibility, technology, and economics of creating a cable system in Washington, D.C., it is full of information that is widely applicable to the creation of cable systems in other urban centers.

Write: Mitre Corporation, 1820 Dolley Madison Blvd., McLean, Va. 22101; or call: 703–893–3500.

Molenda, Michael. *Annotated Bibliography on the Educational Implications of Cable Television (CATV) and Video Cassettes.* n.p.: National Academy of Education, October 1971. Mimeographed, 24pp.

Another excellent bibliography.

Write: Michael H. Molenda, Assistant Professor, School of Education, University of North Carolina, Greensboro, N. Car. 27412.

Price, Monroe and John Wicklein. *Cable Television: A Guide to Citizen Action*. New York: United Church of Christ Press, 1972. 200pp. $2.95.

One copy will be sent free to each community. Write: Office of Communications, United Church of Christ, 289 Park Ave. South, N.Y., N.Y. 10010; or call: 212–475–2127.

Publi-Cable. *Position Paper on Public Issues in Cable Communications.*

Released in December 1971 in mimeographed form, this will be available in printed form by the time this book appears. Publi-Cable is a consortium of representatives of more than forty U.S. civic, cultural, religious, and community groups interested in cable. Its views are interesting and important.

Write: Publi-Cable, Inc., 1201 Sixteenth St., N.W., Washington, D.C. 20036; or call: 202-833-4120.

Rand Corporation. Three studies have been issued by Rand since Shafer compiled his bibliography, that deserve notice. They are:

Baer, W. S. *Prospects for Two-Way Services on Cable*. R-888-MF. November 1971.

Johnson, Leland L. *Cable Television and Higher Education: Two Contrasting Experiences*. R-828-MF. September 1971.

Mitchell, Michael R. *State Regulation of Cable Television*. R-783-MF. October 1971.

Sloan Commission on Cable Communications. *On the Cable: The Television of Abundance*. New York: McGraw-Hill, 1971. 256pp. $2.95.

Book-length report by a blue-ribbon commission set up by the Sloan Foundation to make a year-long study of all aspects of cable TV and broadband communications.

Available in commercial bookstores.

Tate, Charles, editor. *Cable Television in the Cities: Community Control, Public Access, and Minority Ownership*. Washington, D.C.: Urban Institute, 1972. 184 pp. $3.95.

An excellent handbook; indispensable for community groups interested in cable.

Write: The Urban Institute, 2100 M St., N.W., Washington, D.C.; or call: 202–223–1950.

United Church of Christ. *A Short Course in Cable*. New York, 1971. 12pp.

An introductory primer on cable, especially designed for use by local civic and community groups. Individual copies free.

Write: Office of Communications, United Church of Christ, 289 Park Ave. South, New York, N.Y. 10010; or call: 212–475–2127.

APPENDIX B

The Perils of Bidding for a Franchise

At FCC's panel hearings on cable TV held in March 1971, one of the persons who testified was Frederick Rebman, executive vice president of Community Television, the educational TV station in Jacksonville, Florida. Among other things, Mr. Rebman described a telephone call that he received when Community Television was one of the bidders for the Jacksonville cable franchise. As soon as the conversation was completed, Rebman wrote it down to the best of his recollection, and at the hearings he read what he had written down. Anyone who thinks that the game of cable franchising in the cities is not played for keeps, should read Mr. Rebman's testimony as set forth below.

EXCERPT FROM TESTIMONY OF
FREDERICK REBMAN, EXECUTIVE VICE PRESIDENT
COMMUNITY TELEVISION, JACKSONVILLE, FLA.
AT FCC CABLE HEARINGS, MARCH 15, 1971

MR. REBMAN: As you know, sir, I made what I hope was an emotional plea for the Commission to be the sole licensing agent and not to pass the responsibility down to the cities in any measure at all. I had a telephone call that I could relate to you, if you so desired, which would present my concern for what I ask you about ownership. I don't know if time permits for that. I deleted it.

CHAIRMAN BURCH: Go ahead.

MR. REBMAN: This is a telephone call I received just before the awarding of the Jacksonville franchise, which was awarded by a lame duck administration going out of office and since indicted and some of them put in prison.

The phone call I am going to relate for you is not verbatim. However, I did write it down. I had some fear for my life, actually, after getting it. I did write it down to the best of my recollection exactly as it occurred.

"CALLER: Mr. Rebman, this is Mr. Smtih. I understand you folks are going into the cable television business in Jacksonville.

"REBMAN: Yes, we are.

"CALLER: We are one of the biggest in the business and would like to go in with you. Are you interested?

"REBMAN: No, we have had several offers to do that and feel we will go alone.

"CALLER: You guys really don't have any money and probably don't even realize what it will take.

"REBMAN: Yes, it will [take] $15 million to $20 million and we have a commitment from a New York bank to loan all of it to us.

"CALLER: I see, but we have better luck persuading politicians who will ultimately award the franchise.

"REBMAN: Then why are you calling me?

"CALLER: Because we want to make this thing clean and help you out.

"REBMAN: Not so, my friend. It is because you think we are going to win.

"CALLER: We don't want a fight, but I will admit you have a chance.

"REBMAN: Right, the chance lies in 2,000 people willing to take the petitions out and no matter who you own in city government you can't fight a referendum.

"CALLER: You are a gutty little bastard. Do you know I place more millions in one day than you are old.

"REBMAN: You must be a very lucky man.

"CALLER: Why don't you manage the system for us? We will pay you $50,000 a year for the rest of your life and you don't even have to show up for work. How does that sound?

"REBMAN: It sounds [as though,] after the franchise award the rest of my life might not take too long.

"CALLER: (Laughter). You are great. I tell you what. We will give you the money at no interest and build the whole damned thing for you under one condition only.

"REBMAN: What is that?

"CALLER: Well, we are building north from Miami and south from New York. When we get to that great big county of yours, for the big interconnect, we will need a real good friend.

"REBMAN: Thanks anyway, but I enjoy doing difficult things. It keeps me out of trouble.

"CALLER: You are great talking to the old man like that. I will be calling you again."

There was more to the call which I omitted for brevity, but there was a clear indication that the caller had had access to information in my office on a continuing basis. We made many references to private information that had not gone to my Board of Trustees.

The next day a man came by who had had a contract to service parts of our building whereby he needed keys to the building. He had that contract for some seven months before and had been doing a superb service within the building.

He said, "I would like to go and have a cup of coffee with you."

I said, "I am busy. If you don't mind, I can tell you the price is right and the service is great."

He said, "No, it is more important than that."

I said, "We can have the coffee here and I can doodle a little bit."

He said, "No, let's take a stroll."

So we took a stroll toward a neighborhood coffee shop at which time he said, "I hear you got a call from the family the other day."

I mulled the man's name around in my head. I said, "I don't really recall anyone with your name calling me."

He said, "Not close family, sort of an uncle."

Then he referred to the call that I related to you almost verbatim and he said, "You were very fortunate. The old man really liked you. You were a fool to turn it down because you were talking to the big man."

That, gentlemen, is just one of the reasons why I am tremendously interested in this Commission being the sole agent to license cable television and also taking a good, hearty look into who presently owns it, because there are a lot of fake cable operators out there that are fronts.

If you look beyond programming itself and realize that cable operators have access to junction boxes which are also used by AT&T, which means that in any city, the cable operator has access to telephone lines for city government, for bankers doing business, and who is to say whether they are repairing lines or tapping lines.

I think it is an issue of immense concern to a number of us in Jacksonville after that experience, and one that should be to the Commission.

APPENDIX C

The Case for Nonprofit Ownership

Edward P. Curtis, Jr., Chairman of the Board of Trustees of the Rochester, N.Y. Area Educational Television Association, urged consideration of nonprofit ownership of cable systems in his testimony at FCC's March 1971 hearings. His testimony on the possible uses of such a cable system in the Rochester and Monroe County areas is reproduced below. Civic and community groups in other towns and cities may wish to study it.

EXCERPT FROM TESTIMONY OF
EDWARD P. CURTIS, JR., CHAIRMAN OF THE BOARD OF TRUSTEES
ROCHESTER AREA EDUCATIONAL TELEVISION ASSOCIATION
AT FCC CABLE HEARINGS, MARCH 16, 1971

MR. CURTIS: Mr. Chairman and members of the Commission:

In Rochester on December 30, 1969, a lame-duck city council with three days remaining in its term of office awarded a CATV franchise to a group of private entrepreneurs. The award was granted in unseemly haste, over the vigorous protest of many community groups, and in specific contravention of ground rules which the administration then in office had established for considering such an award.

The group receiving the franchise has had a shifting pattern of ownership over the past 15 months, but it looks as if control is now vested in Time-Life Inc. No visible progress has been made in implementing the franchise, and the award triggered two lawsuits which are still waiting trial.

As we look across the nation, this manner of awarding franchises seems to be more and more the norm rather than the exception. Scandal, indictment, corruption, and chicanery mark the rush to give away control over the most promising breakthrough in communications technology since the advent of broadcast TV.

We at RAETA have been delighted to work with community and governmental leaders in Monroe County to try to devise a system that will truly benefit the public interest and insure the broadest possible use of this medium. Specifically, the Monroe County Legislature, in a resolution passed in August, 1970, requested the New York State Legislature to grant it authority to "create a non-profit publicly controlled body to construct and operate a CATV system within Monroe County"—a body that would "represent the public interest to the greatest possible degree, by assuring adequate representation of City and County Government, local educational systems, community leadership and local art and cultural institutions; by designing a system that will be technically capable of carrying local and distant commercial programming and serving as a two-way communication facility for local health, education, public safety and public works authorities; and by an equitable system of distributing CATV revenues which will benefit local governmental, education and cultural institutions."

What advantages do we see in this kind of a system? Those who oppose the licensing of cable television systems to non-profit organizations created by local government make much of the virtues of the competitive free enterprise system. They cling to this view despite the mournful history of radio and television.

We urge an approach to CATV that we believe would be truly competitive—and truly in keeping with the needs and desires of our American institutions. Let there be competition to see what kind of CATV can best meet the needs of our schools and colleges, our hospitals and museums, our cities and our towns. Let there be competition to see what kind of cable television system can best meet the increasing demands of our minorities and others for a fair degree of access to the means of public communications. Let there be competition to determine who has the imagination to develop CATV as a means of making substantial improvements in our way of life—in working toward the solution to our many problems.

Surely the record of privately-owned radio and TV is a discouraging memorial to the glories of free enterprise. We are blessed with a commercially-oriented mass communications system that is the envy of no other major nation in the world, and one that is barely tolerable to those who operate it. So we see no special merit in private ownership, per se.

I think our most important reservation, however, is that we fear the limitations automatically and inescapably imposed on a system designed to be run for profit. Let's face it. The management of any enterprise that is ultimately responsible to investors seeking a profit is bound to operate so as to maximize profit for those investors.

There is no question that short-term profit maximization is best realized by building low-capacity systems that can penetrate the market rapidly for a much smaller investment than a system designed at the cutting edge of present technology.

Specifically, to cite a figure often quoted, in 1969 only 26 out of 2,300 cable systems in the country carried more than 12 channels. A technical and financial feasibility study that we have had conducted for Monroe County calls for a 32-channel system that would serve 95 percent of the residents of Monroe County with a 12-channel return system tied into key public safety, health, education and community centers.

This expanded technology is vital if we are to even begin to tap the resources that are available in the Rochester community for first-class local programming.

Metropolitan Rochester is a community richly endowed with educational, artistic and cultural institutions, many of which are capable today of producing superbly high quality programs that would be ideally suitable for television presentation.

The Rochester Museum and Science Center (with its associated Strasenburgh Planetarium), The University of Rochester's Memorial Art Gallery and Eastman School of Music, the unique collection of classic films at Eastman House, The Rochester Philharmonic Orchestra, the Rochester Public Library—these are but a small sampling of the cultural resources of our community.

Beyond the cultural field, there is an overriding need to return to a sense of local communities in our major cities today. At last count Rochester had 103 neighborhood associations which have been established to preserve or improve the quality of life in their neighborhoods.

The concerns they express and the power they are beginning to wield is a heartening example of genuine citizen participation in what is happening in their communities. This trend needs the kind of strengthening that a community-oriented CATV system can bring.

Providing air time is only half the battle. We can offer channels to the schools and universities, the art galleries and museums, minority groups and neighborhood associations—but unless they have the technical and especially the financial support necessary to underwrite their efforts, they will not be able to develop the full potential of these opportunities.

One of the truly exciting aspects of the Monroe County Legislature's proposal is the possibility of assigning significant amounts of the revenue generated by CATV to the greater development of top-notch public service programming.

In this opening statement, I have tried to avoid conceptual generalities and bring this discussion down to the level of one community's needs. I am sure my story can be repeated again and again across the land—I think it illustrates why the FCC must be sensitive and responsive to the question of public ownership of CATV facilities.

INDEX

73 74 12 11 10 9 8 7 6 5 4 3 2